AF839772

Identification of Heterotrophic Bacteria that Colonize Chloraminated Drinking Water Distribution Systems

Awwa
Research
Foundation

About the Awwa Research Foundation

The Awwa Research Foundation (AwwaRF) is a member-supported, international, nonprofit organization that sponsors research to enable water utilities, public health agencies, and other professionals to provide safe and affordable drinking water to consumers.

The Foundation's mission is to advance the science of water to improve the quality of life. To achieve this mission, the Foundation sponsors studies on all aspects of drinking water, including supply and resources, treatment, monitoring and analysis, distribution, management, and health effects. Funding for research is provided primarily by subscription payments from approximately 1,000 utilities, consulting firms, and manufacturers in North America and abroad. Additional funding comes from collaborative partnerships with other national and international organizations, allowing for resources to be leveraged, expertise to be shared, and broad-based knowledge to be developed and disseminated. Government funding serves as a third source of research dollars.

From its headquarters in Denver, Colorado, the Foundation's staff directs and supports the efforts of more than 800 volunteers who serve on the board of trustees and various committees. These volunteers represent many facets of the water industry, and contribute their expertise to select and monitor research studies that benefit the entire drinking water community.

The results of research are disseminated through a number of channels, including reports, the Web site, conferences, and periodicals.

For subscribers, the Foundation serves as a cooperative program in which water suppliers unite to pool their resources. By applying Foundation research findings, these water suppliers can save substantial costs and stay on the leading edge of drinking water science and technology. Since its inception, AwwaRF has supplied the water community with more than $300 million in applied research.

More information about the Foundation and how to become a subscriber is available on the Web at **www.awwarf.org.**

Identification of Heterotrophic Bacteria that Colonize Chloraminated Drinking Water Distribution Systems

Prepared by:
Daniel R. Noguera, **L. Safak Yilmaz**, and **Gregory W. Harrington**
Department of Civil and Environmental Engineering
University of Wisconsin-Madison, 1415 Engineering Drive, Madison, WI 53706
and
Ramesh K. Goel
Department of Civil and Environmental Engineering
University of Utah, Salt Lake City, UT 84112

Sponsored by:
Awwa Research Foundation
6666 West Quincy Avenue, Denver, CO 80235-3098

Published by:

Awwa Research Foundation

Distributed by:

American Water Works Association

IWA Publishing

DISCLAIMER

This study was funded by the Awwa Research Foundation (AwwaRF). AwwaRF assumes no responsibility for the content of the research study reported in this publication or for the opinions or statements of fact expressed in the report. The mention of trade names for commercial products does not represent or imply the approval or endorsement of AwwaRF. This report is presented solely for informational purposes.

Copyright © 2008
by Awwa Research Foundation

ALL RIGHTS RESERVED.
No part of this publication may be copied, reproduced
or otherwise utilized without permission.

CONTENTS

TABLES

FIGURES

FOREWORD

The Awwa Research Foundation is a nonprofit corporation that is dedicated to the implementation of a research effort to help utilities respond to regulatory requirements and traditional high-priority concems of the industry. The research agenda is developed through a process of consultation with subscribers and drinking water professionals. Under the umbrella of a Strategic Research Plan, the Research Advisory Council prioritizes the suggested projects based upon current and future needs, applicability, and past work; the recommendations are forwarded to the Board of Trustees for final selection. The foundation also sponsors research projects through the unsolicited proposal process; the Collaborative Research, Research Applications, and Tailored Collaboration programs; and various joint research efforts with organizations such as the U.S. Environmental Protection Agency, the U.S. Bureau of Reclamation, and the Association of Califomia Water Agencies.

This publication is a result of one of these sponsored studies, and it is hoped that its findings will be applied in communities throughout the world. The following report serves not only as a means of communicating the results of the water industry's centralized research program but also as a tool to enlist the further support of the nonmember utilities and individuals.

Projects are managed closely from their inception to the final report by the foundation's staff and large cadre of volunteers who willingly contribute their time and expertise. The foundation serves a planning and management function and awards contracts to other institutions such as water utilities, universities, and engineering firms. The funding for this research effort comes primarily from the Subscription Program, through which water utilities subscribe to the research program and make an annual payment proportionate to the volume of water they deliver and consultants and manufacturers subscribe based on their annual billings. The program offers a cost-effective and fair method for funding research in the public interest.

A broad spectrum of water supply issues is addressed by the foundation's research agenda: resources, treatment and operations, distribution and storage, water quality and analysis, toxicology, economics, and management. The ultimate purpose of the coordinated effort is to assist water suppliers to provide the highest possible quality of water economically and reliably. The true benefits are realized when the results are implemented at the utility level. The foundation's trustees are pleased to offer this publication as a contribution toward that end.

David E. Rager
Chair, Board of Trustees
Awwa Research Foundation

Robert C. Renner, P.E.
Executive Director
Awwa Research Foundation

ACKNOWLEDGMENTS

The authors gratefully acknowledge the contributions of numerous individuals and organizations in this effort. The Louisville Water Company provided water samples from their distribution system and from a pilot-scale distribution system simulator. Special thanks to Rengao Song, Manager of Water Quality and Research, Mark Campbell, Pilot Plant Engineer, Roger Tucker, Research Scientist, and Por, Pilot Plant Operator. Milwaukee Water Works also provided samples from the distribution system. Special thanks to Lon A. Couillard, Water Quality Manager for his support and helpful suggestions.

During the course of this project, the pilot plant at the University of Wisconsin was operated by Prapakorn Assavasilavasukul and Soumya Srinivasan. Several undergraduate students contributed to the operation of the bench-scale reactors, including Yussuf Jamaa and Kaitlin Krause. The assistance of Jackie Bastyr Cooper, Manager of the Environmental Engineering Laboratory is also greatly appreciated. Administrative support at the University of Wisconsin was provided by Darlene Sayre, Ken Ostrander, Tamara Kuhn Martin, Vicki Johnson, and Mary Koscielniak.

The authors also thank John Albert, Awwa Research Foundation Project Manager, for his help and support throughout the project. The support and suggestions from the project advisory committee are also gratefully acknowledged. The members of the project advisory committee were Christian Chauret (Indiana University - Kokomo), Susan Springthorpe (University of Ottawa), and Moyasar Yahya (Charles Abbott Associates, Mission Viejo, CA).

EXECUTIVE SUMMARY

BACKGROUND

The problem of bacterial regrowth in chloraminated drinking water distribution systems has been extensively studied under the assumption that ammonia-oxidizing bacteria (AOB) are the microorganisms responsible for initiating the deterioration of water quality in the system (Wolfe et al. 1990; Cunliffe 1991; Lieu et al. 1993; Oldenburg et al. 2002; Regan et al. 2002; Pintar and Slawson 2003; Regan et al. 2003). Under this conceptual scheme, the subsequent regrowth of heterotrophic bacteria can be explained by the dramatic decrease in disinfectant concentration after the onset of nitrification, which is exacerbated by the presence of a higher concentration of microorganisms, the development of a biofilm, and the chemical reactions between nitrite and chloramines. However, in some cases heterotrophic growth is observed before evidence of nitrification is detected, which could indicate a different scenario of interaction between heterotrophs and nitrifiers within chloraminated distributions systems. For instance, disinfectant-resistant heterotrophs could initially colonize a distribution system if biodegradable dissolved organic carbon (BDOC) is present. Heterotrophic bacteria growth will then promote more rapid chloramine decomposition within the distribution system, which would release ammonia and create the conditions necessary for the onset of nitrification. Nitrification control strategies for this second scenario would not need to focus on controlling AOB growth, but rather on minimizing the growth of disinfectant-resistant heterotrophs.

In the past, microbial ecology studies have focused on identifying the prevalent AOB and nitrite-oxidizing bacteria (NOB) in chloraminated systems (Regan et al. 2002; Regan et al. 2003), and have demonstrated that *Nitrosomonas oligotropha* and *Nitrospira spp*. are ubiquitous AOB and NOB, respectively, in the drinking water environment. Information regarding the identity of heterotrophic bacteria in chloraminated distribution systems is scarce, with only one major study that compared 16S rRNA sequences from chlorinated and chloraminated distribution system simulators (Williams et al. 2004; Williams et al. 2005) and one lab-scale study that identified a limited number of microorganisms grown in annular reactors with different disinfectants including chloramines (Gagnon et al. 2007). As a result of this gap in knowledge potential strategies to control nitrification in chloraminated distribution systems by reducing the growth of heterotrophic bacteria that are potentially the first colonizers of the distribution systems have not been developed. As a first step in this direction, this project aimed at filling the gap in knowledge regarding the identity of heterotrophic bacteria that grow in chloraminated distribution systems.

OBJECTIVES

The research described in this report focused on the following objectives:

1. Identifying the heterotrophic bacteria that colonize chloraminated drinking water distribution systems using bench-scale, pilot-scale, and full-scale chloraminated environments.
2. Determining whether bench-scale and pilot-scale chloraminated systems are adequate models to study the type of heterotrophic bacteria that colonize full-scale chloraminated distribution systems.

3. Exploring the development of a rapid screening test, based on the differences in length of the internal transcribed spacer (ITS) region of the rRNA bacterial operon, to identify culturable heterotrophic bacteria in drinking water distribution systems.

GENERAL APPROACH

Nine environments were studied and included the following:

1. Two bench-scale reactors simulating chloraminated and non-chloraminated environments, respectively;
2. Two pilot-scale distribution system simulators that received water treated in a pilot-scale treatment plant operated at the University of Wisconsin, with Lake Mendota (Madison, WI) as the source of water. The chloramine concentration in one distribution system simulator was targeted to be between 0.5 and 1.0 $mgCl_2/L$, while the second distribution system was targeted to have concentrations between 1.5 and 2.0 $mgCl_2/L$.
3. Two pilot-scale distribution system simulators operated by the Louisville Water Company (Louisville, KY) and received finished water from the full-scale treatment plant. One simulator was not chloraminated while the second simulator was chloraminated and also received chlorite.
4. Finished water from the Louisville Water Company treatment plant.
5. Two water tanks in the distribution system at Louisville (Prospect Tank and Martin's Hill Tank).

For all environments, only bulk water samples were analyzed. Samples from biofilms were not collected due to funding limitations.

A molecular biology approach based on the 16S rRNA as a phylogenetic marker was used to identify and characterize the heterotrophic communities in all samples. In addition, the variable length of the ITS region in the rRNA operon, which is common to all bacteria, was used for community fingerprinting using the automated ribosomal intergenic spacer analysis (ARISA) (Fisher and Triplett 1999), which allowed the comparison of microbial communities among the different chloraminated environments tested. Finally, the ITS variability was also explored as a rapid screening and identification test for colonies grown in agar plates.

RESULTS

Analysis of 16S rRNA gene sequences

A library of 16S sequences was generated from water samples collected from the bench-, pilot-, and full-scale environments studied. Overall, a total of 1,371 partial sequences and 197 full sequences of the 16S rRNA gene were obtained. These sequences were analyzed using three approaches:

1. For each tested environment, sequences were arranged according to their taxonomic hierarchy. This was useful to identify what phylogenetic groups were present and their frequency within the library.
2. Sequences in the library were compared to publicly available databases to identify instances when closely matching organisms have been observed before. Through this level of analysis, we extracted from the databases information related to the

environments where the closely matching organisms were identified. Fifty five of the partial 16S rRNA sequences obtained in this project were virtually identical to published sequences in the Ribosomal Database Project (RDP) database (Cole et al. 2005).

3. The library of sequences was organized into operational taxonomic units (OTU) according to their similarity to each other. Representatives of each OTU were then selected for the construction of phylogenetic trees that allowed the visualization of relatedness of the defined OTUs to each other, to relevant public sequences, and to well-defined phylogenetic groups as well as to uncultured bacteria.

The taxonomic classification of all partial sequences in the library revealed the following features:

1. At the phylum level, all nine environments tested included *Actinobacteria*, *Bacteroidetes*, and *Proteobacteria*, and in general, these three phyla represented the majority of the sequences in the library. *Nitrospira* and *Firmicutes* were detected in 5 out of the 9 environments tested. *Spirochaetes* were identified in 2 environments, while *Acidobacteria* were detected in only one environment. Overall, microorganisms belonging to only 7 phyla out of the 33 phyla currently recognized in the domain Bacteria were detected.
2. Among *Actinobacteria*, *Mycobacterium* were found in all systems, with remarkable frequency in all the pilot-scale distribution system samples and in the samples from the Martin's Hill Tank. The members of this genus have significant relevance to drinking water treatment, as they are notoriously resistant to disinfection, preferentially form biofilms, and some are known pathogens.
3. Most sequences under the phylum *Bacteroidetes* were assigned to the *Flavobacteriales* and *Sphingobacteriales* families, and a lower number of sequences to the *Bacteroidales* family. All sequences related to the *Bacteroidales* family belonged to the *Dysgonomonas* genus and were only detected in samples received from Louisville (pilot-scale and full-scale samples), so their presence may have a particular geographical meaning or could be representative of the source of water.
4. Sequences under the phylum *Firmicutes* were particularly prevalent on samples from the full-scale system, with *Bacilli* and *Mollicutes* being the representative classes. Among the *Bacilli* class, members of the genus *Staphylococcus*, *Dolosigranulum*, and *Streptococcus* were recurrent. All sequences within the *Mollicutes* were classified in the *Erysipelotrichaceae* family, but were not classified in any specific genus.
5. Sequences assigned to the *Nitrospira* phylum were found in bench-scale, pilot-scale, and full-scale samples, but were a minor component of the library. In all cases, clones were classified under the *Nitrospira* genus, which includes nitrite oxidizing bacteria, and therefore are indicators of nitrification in these systems.
6. By far, the majority of the sequences belonged to the phylum *Proteobacteria*. Sequences were generally distributed among the *Alpha*-, *Beta*-, and *Gammaproteobacteria*, indicating a large share of these classes consistent with other clone libraries of drinking water (Williams et al. 2004; Tokajian et al. 2005). Among the *Alphaproteobacteria*, members of the orders *Caulobacterales*, *Rhizobiales*, and *Sphingomonadales* appeared to be significant in most samples. *Betaproteobacteria* included clones belonging to the *Nitrosomonas* genus associated with ammonia oxidation. Most of the other *Betaproteobacteria* were classified under the

Burkholderiales order, with the *Limnobacter* genus being a frequent observation. The presence of *Gammaproteobacteria* was also common. The presence of *Legionella* and *Pseudomonas* related sequences is also noted.

7. The *Sphingomonadales* order appeared to be particularly associated with samples from chloraminated environments, as the only environments in which they were not detected were the finished water from Louisville and the non-chloraminated bench-scale reactor.

Analysis of microbial community profiles

In addition to the identification of microorganisms by 16S rRNA gene sequences, the bacterial community in the studied environments was characterized using the automated ribosomal intergenic spacer analysis (ARISA) approach. The main results from the ARISA analyses are the following:

1. A comparison of the community ARISA profiles for the bench-scale reactors demonstrated that the microbial communities enriched in each reactor were not the same, in agreement with the operational mode of only one reactor receiving chloramines. It was possible to identify specific ARISA peaks that were present in the chloraminated reactor but not in the non-chloraminated reactor, and vice versa. It was also possible to identify ARISA peaks that were common to both reactors. With the matching of ARISA peaks to 16S sequencing, a potential identification of members of the community that were common to both reactors and community members that were specific to each of the reactors was possible, but should not be considered as an unequivocal identification.
2. The clustering of all pilot-scale ARISA profiles revealed that the bacterial communities in the Louisville pilot plants were significantly different from the communities observed in the UW pilot plants. Moreover, the communities in the two simulators at Louisville were not significantly different from each other.
3. The clustering analysis of ARISA profiles from the two UW plants showed that an important parameter was the period of sampling. Common organisms in both pilot-scale distribution systems included members of the *Mycobacterium* genus and members of the *Sphingomonadaceae* family within the *Alphaproteobacteria*. Organisms that were common to all pilot-scale systems, including those operated at Louisville could be matched to the genera *Mycobacterium*, *Sphingomonas*, *Hypobacterium*, and *Porphyrobacter*.
4. Unlike bench- and pilot-scale samples, the clustering analysis of full-scale samples revealed no clear subclusters of individual sampled locations. As such, the microbial communities in the Martin's Hill Tank and and the Prospect Tank appeared to have significant overlap. A general observation was the frequent appearance of ARISA peaks potentially related to *Mycobacterium*, the Sphingomonadales order of the *Alphaproteobacteria* and the *Sphingobacteriales* order of the *Bacteriodetes*, which appear to be common to all environments studied, regardless of the scale. Other peaks in the ARISA profiles of full-scale samples could be associated with *Stenotrophomonas*, *Pseudomonas*, *Hyphomicrobium*, and *Methylobacterium*.

A particular objective of this study was to evaluate whether chloraminated bench- and pilot-scale systems could represent full-scale environments with regard to bacterial community composition. As discussed above, ARISA fingerprints showed statistically significant differences between the scales, with some overlap between pilot- and full-scale. In addition, phylogenetic trees also showed interesting relationships between sequences retrieved from the different scales. Each phylogenetic group identified included sequences from all three scales, suggesting that the microbial communities of the simulators and full-scale systems are similar at higher levels of taxa. However, at lower orders of clustering, there were clusters of mixed scales suggesting bench-scale chemostats and pilot-scale simulators were able to grow organisms inhabiting full-scale distribution systems, including but not limited to the *Sphingomonadales* family, and the *Pseudomonas, Limnobacter,* and *Mycobacterium* genera. Nevertheless, there were also lineages involving only full-scale distribution system clones such those representing the *Azonexus* and *Staphylococcus* genera. The opposite was also observed. For example, the *Arthrobacter* genus was represented by only bench-scale chemostat clones, suggesting that *Arthrobacter*-related species may not be relevant to full-scale systems. These observations imply that bench- and pilot-scale systems may introduce biases in the simulation of distribution systems. However, another possible explanation for the presence of organisms not common to all scales is the influence of system-specific factors such as the source water, which might have affected the microbial community composition due to differences in water quality or the microbial seed. In this study, systems at the three different scales had different source waters.

Thus, controlled reactors and simulators, as those sampled in this study, may be used to grow and investigate types of organisms inhabiting real scale distribution systems, but representative ability at the species level should be handled with care. Finally, it was also evident in this study that bench-scale reactors were more prone to contamination by other organisms present in the laboratory, which makes them less ideal for meaningful community composition studies.

Evaluation of ARISA as a tool for identification of culturable microorganisms from drinking water distribution systems

An additional molecular level database generated in this study was the library of ARISA peaks for most of the recovered clones, which represented the variability in ITS fragment lengths. Initially, the clone-ARISA database was expected to be useful in tracking different types of organisms in community profiles without need for additional sequencing. However, the predictions of peak identities were not unequivocal since it was determined that one peak could be representing several microorganisms from different phylogenetic groups. This problem is accentuated by the presence of multiple rRNA operons in a typical bacterial cell, which means that a single organism can be represented in an ARISA profile by more than one peak. Thus, with the large diversity of bacterial species and the multiciplicity of rRNA operons, there variability of the ITS length is not sufficient to allow unequivocal assignments of peak length to bacterial identity. Nonetheless, ARISA fingerprints were still useful for comparative statistical analysis of community profiles, and the probable matching of alleles in these fingerprints to 16S identity generally confirmed the findings from clone libraries.

RECOMMENDATIONS

For utilities interested in exploring the role of heterotrophs in the onset of nitrification in chloraminated systems, it is recommended that they develop a semi-quantitative understanding of whether nitrifiers are the first colonizers in their distribution system or whether chloramine-resistant heterotrophs initiate water deterioration that results in nitrification. For this, general measurements of heterotrophic and nitrifier growth, such as heterotrophic plate counts and most probable number determinations could be useful to assess the relative growth of these two microbial groups within the distribution system.

For utilities interested in identifying the type of heterotrophic bacteria growing in the distribution systems, cloning and sequencing the 16S rRNA gene is the most direct approach. However, due to the cost of these analyses, the information provided by 16S sequencing is finite and dependent on the number of sequences afforded, and therefore, it is recommended that other methods be used for community profiling along with 16S sequencing. In this report, we demonstrate the usefulness of ARISA as a community fingerprinting tool. Moreover, if a utility decides to use 16S rRNA sequencing for bacterial identification, it is recommended the cloning and amplification of not only the 16S rRNA gene, but also of the ITS region. That way, the full benefit of matching ARISA peaks with bacterial identity could be realized.

Finally, the matching of ARISA peaks with 16S sequence information was demonstrated to be useful for the analysis of bacterial communities from different environments. The library of matching information presented in this report is incomplete due to the obvious resource limitations. It is recommended that such library continue to be expanded to include a more distributed set of environments, included samples from full-scale systems with a variety of water sources, treatment schemes, disinfection practices, etc. Only then, could the full benefit of the ARISA peak identification initiated in this project be fully realized.

CHAPTER 1
INTRODUCTION

BACKGROUND

The use of chloramines, as an alternative to free chlorine for disinfection of drinking water, has significantly increased in the United States (Wilczak et al. 1996; AWWA 2000) in response to stringent limits on halogenated organic compounds (USEPA 1998) typically formed by the interaction of hypochlorous acid (or hypochlorite) with naturally occurring organic matter and bromide (Brodtmann and Russo 1979). Chloramines also offer the advantage of improving the capacity to maintain a disinfectant residual throughout the distribution system (Norton and LeChevallier 1997). Despite the potential advantages of chloramination, the use of chloramines can also lead to biological instability in the distribution system by promoting the growth of nitrifying bacteria, microorganisms that oxidize ammonia to nitrite and nitrate. The free ammonia[1] introduced to the distribution system during chloramine formation and from chloramine decay can serve as a growth substrate for ammonia-oxidizing bacteria (AOB). As AOB grow and decay, they produce nitrite and organic compounds, which exert a chloramine demand and support the growth of nitrite-oxidizing bacteria (NOB) and heterotrophs. This microbial activity can result in conditions favorable for biological regrowth, potentially leading to violations of the Total Coliform Rule (USEPA 1989), the disinfectant residual requirement of the Surface Water Treatment Rule (USEPA 1989), and in extreme cases, the nitrite maximum contaminant level of 1 mg N/L (USEPA 1991). Evidence of nitrification episodes has been reported by nearly two-thirds of chloraminating utilities (Wilczak et al. 1996), and the consequent increase in heterotrophic growth and difficulty in maintaining residuals is well documented (Wolfe et al. 1990; Cunliffe 1991; Skadsen 1993; Odell et al. 1996; Harrington et al. 2003).

The problem of bacterial regrowth (both nitrifiers and heterotrophs) in chloraminated drinking water distribution systems has been extensively studied under the assumption that AOB are the microorganisms responsible for initiating the deterioration of water quality in the system (Wolfe et al. 1990; Cunliffe 1991; Lieu et al. 1993; Oldenburg et al. 2002; Regan et al. 2002; Pintar and Slawson 2003; Regan et al. 2003). This is a reasonable assumption, since water quality deterioration is mainly associated with the decay of chloramines and the resulting release of ammonia, which is the food and energy source for AOB. Under this conceptual scheme, the subsequent regrowth of heterotrophic bacteria can be explained by the dramatic decrease in disinfectant concentration after the onset of nitrification, which is exacerbated by the presence of a higher concentration of microorganisms, the development of a biofilm, and the chemical reactions between nitrite and chloramines. However, in some cases heterotrophic growth is observed before evidence of nitrification is detected, which could indicate a different scenario of interaction between heterotrophs and nitrifiers within chloraminated distribution systems. For instance, disinfectant-resistant heterotrophs could initially colonize a distribution system if biodegradable dissolved organic carbon (BDOC) is available. Heterotrophic biofilm development will then promote more rapid chloramine decomposition within the distribution system, which would release ammonia and create the conditions necessary for the onset of nitrification. Nitrification control strategies for the latter scenario would not need to focus on

[1] For simplicity, free ammonia in this project is equivalent to the sum of ammonia and ammonium ion.

controlling AOB growth, but rather on minimizing the growth of disinfectant-resistant heterotrophs.

The standard practice for the assessment of microbial regrowth in distributions systems is the generic measurement of heterotrophic bacteria using heterotrophic plate counts (HPC). Furthermore, the growth of nitrifiers is sometimes evaluated using most probable number (MPN) techniques. Both approaches provide information on the abundance of microorganisms in the distribution system, but do not inform on the type of heterotrophs or nitrifiers that colonize these distribution systems. In the past, microbial ecology studies have focused on identifying the prevalent AOB and NOB in chloraminated systems (Regan et al. 2002; Regan et al. 2003), and have demonstrated that *Nitrosomonas oligotropha* and *Nitrospira spp.* are ubiquitous AOB and NOB, respectively, in the drinking water environment. Information regarding the identity of heterotrophic bacteria in chloraminated distribution systems is scarce, with only one major study that compared 16S rRNA sequences from chlorinated and chloraminated distribution systems simulators (Williams et al. 2004) and one lab-scale study that identified a limited number of microorganisms grown in annular reactors with different disinfectants including chloramines as well as those without any treatment (Gagnon et al. 2007). Thus, this project aimed at filling a gap in knowledge regarding the identity of heterotrophic bacteria that grow in chloraminated distribution systems.

OBJECTIVES

The research described in this report focused on the following objectives:

1. Identifying the heterotrophic bacteria that colonize chloraminated drinking water distribution systems using bench-scale, pilot-scale, and full-scale chloraminated environments.
2. Determining whether bench-scale and pilot-scale chloraminated systems are adequate models to study the type of heterotrophic bacteria that colonize full-scale chloraminated distribution systems.
3. Exploring the development of a rapid screening test, based on the differences in length of the inter-transcribed spacer (ITS) region of the rRNA bacterial operon, to identify culturable heterotrophic bacteria in drinking water distribution systems.

GENERAL APPROACH

A molecular biology approach based on the 16S rRNA as a phylogenetic marker (Woese 1987; Olsen and Woese 1993) was used to identify and characterize the heterotrophic communities in chloraminated systems. In addition, the variable length of the ITS region in the rRNA operon, which is common to all bacteria, was used for community fingerprinting using the automated ribosomal intergenic spacer analysis (ARISA) (Fisher and Triplett 1999), which allowed the comparison of microbial communities among the different chloraminated environments tested. Finally, the ITS variability was also explored as a rapid screening and identification test for colonies grown in agar plates.

Although biofilms undoubtedly play a role in defining the microbial communities found in distribution systems, for simplicity and due to funding constraints, only bulk water samples were analyzed.

REPORT ORGANIZATION

This introductory chapter is followed by a detailed description of the distribution systems studied and their performance during the project period. Then, the methodology used in microbial ecological analyses is explained in Chapter 3. The main results and findings from phylogeny and community fingerprints are presented in Chapters 4 and 5, respectively. Chapter 6 outlines and gives proof-of-principle results for the screening and identification of heterotrophic bacteria based on the variability of the ITS region. Finally, a summary of conclusions and recommendations is provided in Chapter 7.

CHAPTER 2
DISTRIBUTION SYSTEMS STUDIED

The studied environments were comprised of chloraminated drinking water distribution systems (DS) at three levels: bench-scale chemostats, pilot-scale DS simulators, and full-scale DS. The comparison of microbial communities in these different environments was implemented to understand if bench- and pilot-scale systems are adequate models of full-scale environments from a microbial ecological point of view, and specifically in relation to the composition of the heterotrophic communities present in chloraminated water experiencing nitrification.

BENCH-SCALE CHEMOSTATS

In bench-scale tests, microbial communities were enriched in two chemostats that received ammonia as the only external electron donor. One of the reactors (Chl+) was chloraminated by mixing the feed with chlorine, whereas the other (Chl-) did not receive any form of disinfectant (Figure 2.1). Both reactors were initially seeded in 2001 with chloraminated water from a full-scale distribution system tank experiencing nitrification problems (Louisville Water Company, KY, USA). Since then they have been operated in a temperature controlled room at 22°C. Prior to this project, these reactors have been used in research aimed at identifying nitrifying bacteria that prevail in chloraminated systems (Regan 2001). Each reactor consisted of a 1L glass vessel with an overflow port, inlet tubing for filter medium, a fritted glass diffuser, and a vent port. While the operational conditions slightly changed during the entire run of the reactors, the explanations provided here pertain to the state in the period of this research (i.e., April 2005 to March 2006).

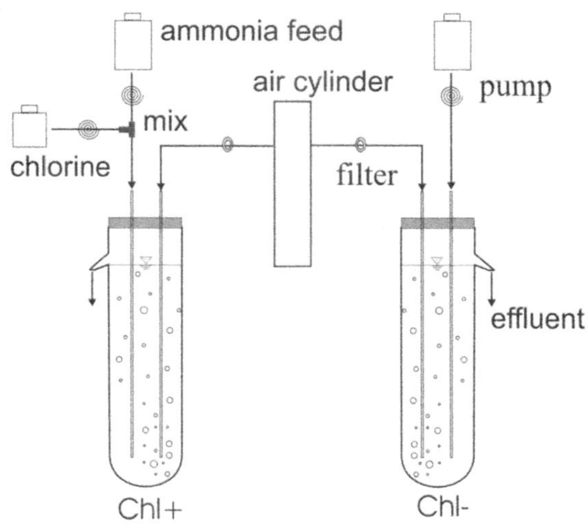

Figure 2.1 Schematic representation of chloraminated (Chl+) and non-chloraminated (Chl-) chemostats

High DO conditions (ca. 8 mg O_2/L) were achieved in both reactors by bubbling air from a compressed air cylinder. Phenol red was used as a pH indicator and the pH was generally maintained between 6.5 and 7.5 in both reactors. The common ingredients of the feed solutions included the following (in mg/L): $CaCl_2$ $2H_2O$, 20; $MgSO_4$ $7H_2O$, 200; K_2HPO_4, 15; $FeSO_4$ $7H_2O$, 0.05; K_2CO_3, 10.4; Na_2MoO_4, 0.1; $MnCl_2$ $4H_2O$, 0.2; $ZnSO_4$ $7H_2O$, 0.1; $CoCl_2$ $6H_2O$, 0.002; $CuSO_4$ $5H_2O$, 0.02. No organic carbon sources were added to the reactors. As nitrogen source, the feeds of Chl- and Chl+ contained 0.07 and 0.1 mM (i.e., 9.3 and 13.2 mg/L) $(NH_4)_2SO_4$, respectively. The medium was introduced to Chl- at 0.1 mL/min to achieve a hydraulic residence time (HRT) of 6.9 days. As for Chl+, the medium was pumped at 0.07 ml/min, but it was premixed with a NaOCl solution provided at 0.03 mL/min, bringing the overall influent rate to 0.1 mL/min, and hence, the same HRT as Chl-.

The concentration of total chlorine in the mixed feed of Chl+ was 3.25 mg Cl_2/L until June 2005. As noted later, nitrification was observed in this chemostat under this operating condition. In an effort to determine if increased chloramine concentrations could change the microbial community, the total chlorine concentration in the mixed feed of Chl+ was increased by stepwise additions of 0.26 mgCl_2/L per month starting July 2005, to reach 5.6 mgCl_2/L in March 2006. These increments were implemented by increasing the concentration of the NaOCl solution. When chlorine is mixed with ammonia, chloramines are formed (Glaze 1990). In the case of a Cl:N ratio (w/w) smaller than 5, the amount of chloramines formed can be estimated from equilibrium chemistry and stoichiometry (Oldenburg et al. 2002; Harrington et al. 2003). The maximum mass ratio in our experiments was 2.9. Calculations following the methods described in Oldenburg et al. (2002) showed that virtually all of the chlorine was converted to chloramines under the pH conditions provided. These calculations also showed that 90% of the combined chlorine was likely to be monochloramine, 10% was likely to be dichloramine, and nitrogen trichloride formation was likely to be negligible.

The specific design of the chemostat experiments served two purposes. First, since no external carbon source was provided, heterotrophic growth was dependent on soluble organic compounds produced by AOB and NOB, as well as soluble organics released during biomass decay. Thus, a microbial community based on the commensalistic relationship between nitrifiers and heterotrophs was aimed to be established. Second, since one of the reactors was chloraminated, the effect of chloramination on the microbial community could be evaluated. With the stepwise increase in chloramines, the experiments aimed at testing whether the microbial community within the Chl+ reactor would change in response to chloramine dose. It should be noted that a gradual increase in disinfectant concentration may allow microorganisms to become resistant by adaptive mutations, and that the final disinfectant concentrations in the influent of the chemostats was higher than used in practice. These two factors may contribute to the development of a different microbial community than observed in pilot- or full-scale distribution systems.

Reactor performance with regard to nitrification was evaluated by routine monitoring of influent and effluent ammonia, and effluent nitrite and nitrate (both were zero in the influent), with colorimetric methods using HACH kits (HACH Company, Loveland, CO). Samples were taken about every two weeks from both reactors. Figure 2.2 depicts time series of the measured species for the duration of the project. For some samples, total chlorine in the effluent of Chl+ was measured using a colorimetric HACH kit, but in all cases the measurement was below detection limit, so it was concluded that the demand in the reactor was higher than the supplied chlorine throughout the duration of the experiments.

During the whole research period, nitrification was evident in both reactors. On average, ammonia concentrations were reduced by 80-90% within the reactors, and reactor effluents contained nitrate concentrations generally comparable to the observed loss of ammonia (Figure 2.2). Exceptions to this were an increase in effluent ammonia coupled with a decrease in effluent nitrate in Chl+ from December 2005 to January 2005, and a temporary accumulation of nitrite in the effluent of the same reactor in February 2005. Therefore, the given environmental conditions were considered as partially favorable for nitrification in Chl+ and as fully favorable in Chl-.

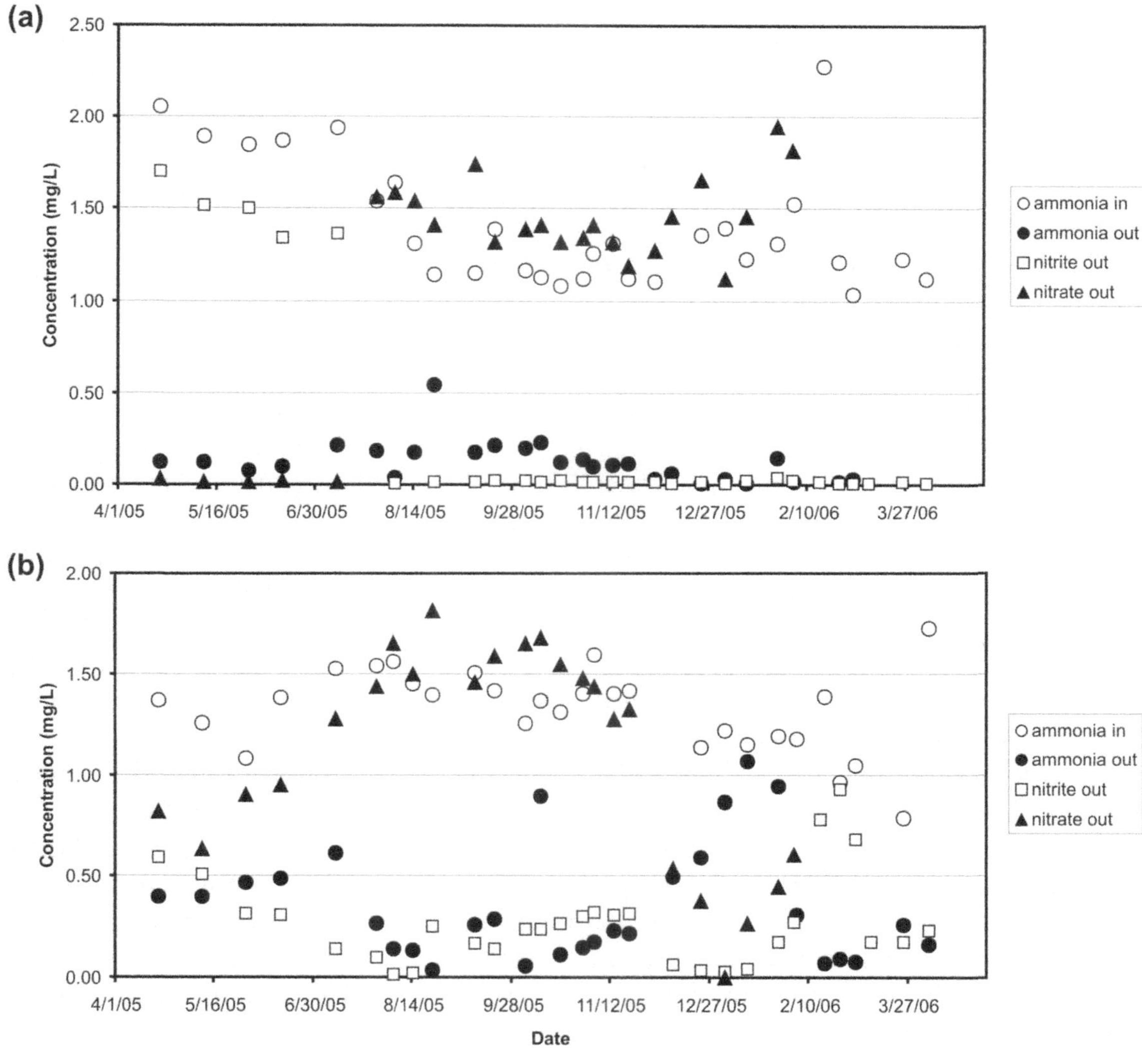

Figure 2.2 Time series of measured nitrogen species in (a) Chl- chemostat and (b) Chl+ chemostat. Concentrations are reported in mg/L as N.

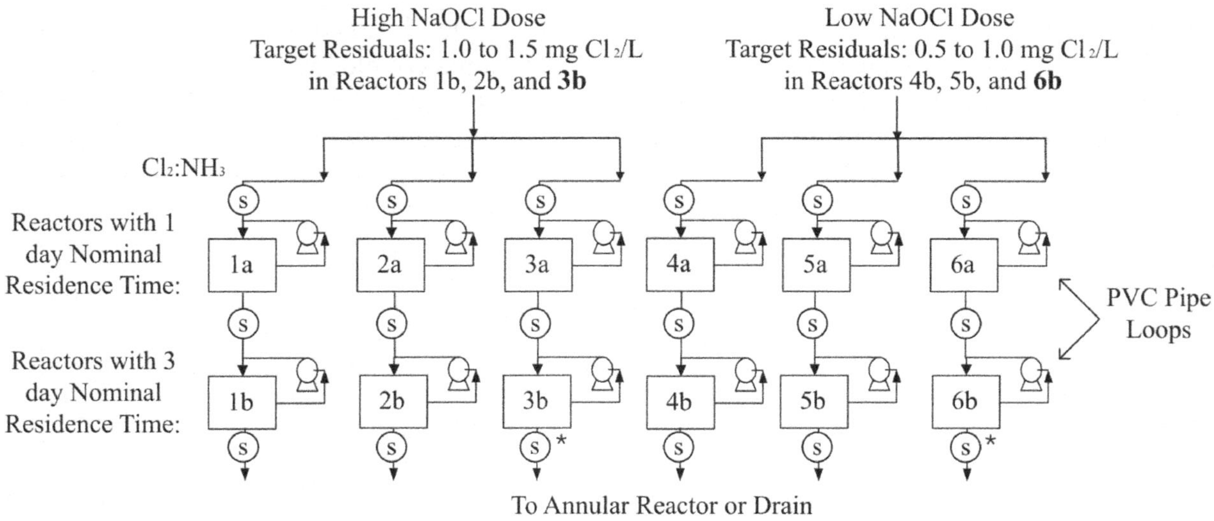

Figure 2.3 Schematic of UW pilot-scale distribution systems

PILOT SCALE DISTRIBUTION SYSTEM SIMULATORS

Pilot-scale distribution systems simulate the fluid motion and environmental conditions typically encountered in full scale distribution systems. At the University of Wisconsin (UW), we operate two pilot systems that treat water from Lake Mendota and have the capability of creating different environmental conditions in separate distribution system loops. Research with these pilot systems has been reported elsewhere (Harrington et al. 2002; Fleming et al. 2005; Yang et al. 2007). For this project, two parallel loops differing in NaOCl dose were operated (Figure 2.3). These loops had a low NaOCl dose [0.5 to 1.0 mg Cl_2/L], designated as UWL; and a high NaOCl dose [1.5 to 2.0 mg Cl_2/L], designated as UWH.. The regular operation of these systems started in June 2005, was temporarily stopped in November 2005, resumed in April 2006, and was operated until January 2007.

A third pilot system was set up at the Louisville Water Company facilities in October 2006 (Figure 2.4). There were four trains in the Louisville pilot system, two of which (LO1 and LO3) were sampled for our analyses. One of these trains was chloraminated (LO3) and the other was not chloraminated (LO1). LO3 was also fed with chlorite to observe its effect on nitrification. The operation and overall design of this pilot system was not part of this project, so information on specific parameters is not reported here.

The Louisville pilot trains were seeded with water from the chloraminated Chl+ reactor defined above, at the end of its operation period.

FULL SCALE DISTRIBUTION SYSTEM

Three sites in the full-scale distribution system at the Louisville Water Company were selected for monitoring in this project. The first site (finished water collected at the plant) represented a location not expected to have nitrification problems or heterotrophic growth. The second site (Prospect Tank) was a location with little nitrification. The third location (Martin's Hill Tank) was intended to evaluate a site with recurrent nitrification problems.

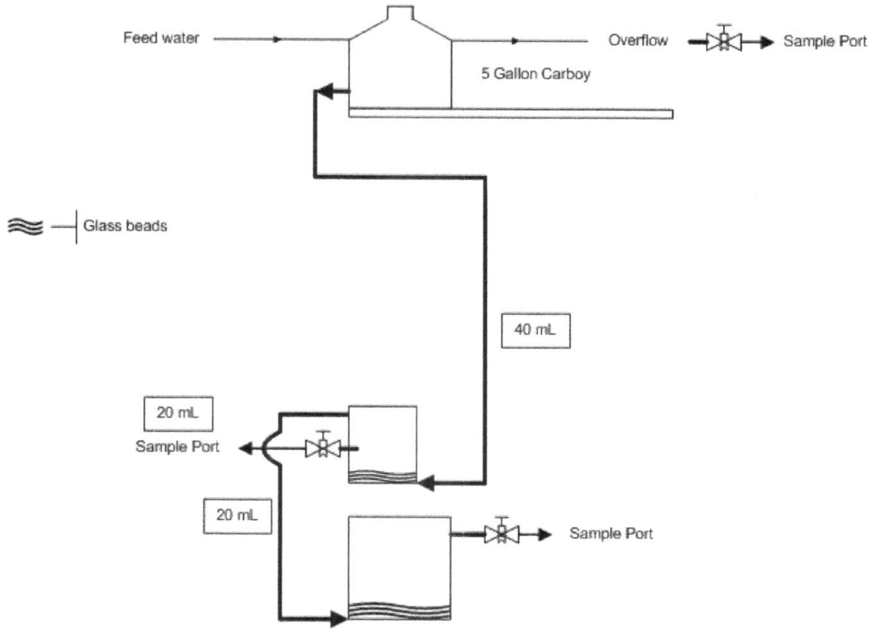

Figure 2.4. Schematic representation of pilot-scale distribution system operated at the Louisville Water Company

CHAPTER 3
METHODOLOGY

OVERVIEW

Methods for identification of bacteria in microbial communities can be classified as culture-dependent and culture-independent. Due to well known biases introduced during culturing of organisms (Amann et al. 1995), culture-independent techniques have been preferred in modern microbial ecological research (Saiki et al. 1988; Amann and Ludwig 2000). Culture-independent identification generally relies on the use of a phylogenetic marker commonly present in all members of the targeted community, but with sufficient variability that can be correlated with taxonomy. The culture-independent techniques also have some biases and these are described in the next section.

In this project, culture-independent techniques were employed for identifying the heterotrohic components of communities colonizing the distribution systems studied. Two phylogenetic markers were used, the 16S ribosomal RNA (rRNA) gene and the internal transcribed spacer (ITS) between the 16S and the 23S rRNA genes, both of which reside in the rRNA operon. This operon is often present in multiple (not identical) copies in the bacterial genome and encodes the information for all rRNA genes (16S, 23S, and 5S) as well as some transfer RNA genes. The bulk of the operon consists of the segment including the 16S and 23S genes and the ITS region between them (Figure 3.1). While the size of the 16S and 23S genes are generally on the order of 1.5 and 2.3 kilobases, respectively, that of ITS varies generally between 100 to 800 bp (Garcia-Martinez et al. 2001).

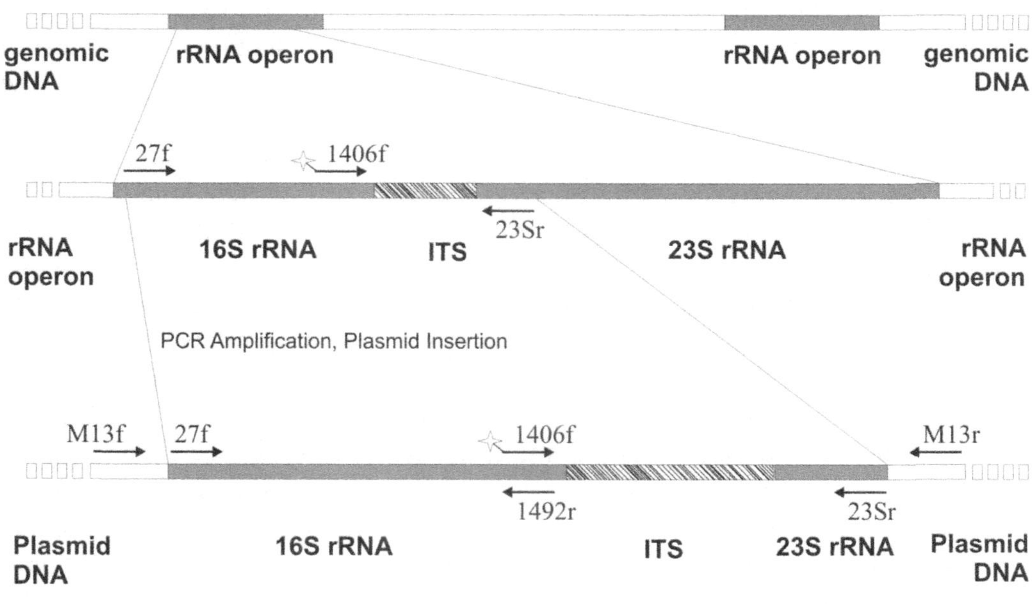

Figure 3.1 Schematic representation of amplified fragments in the rRNA operon. Top panel, rRNA operons on the genome; middle panel, organization of relevant genes in the rRNA operon; bottom panel, plasmid-inserted region. Arrows indicate primer target positions (not to scale).

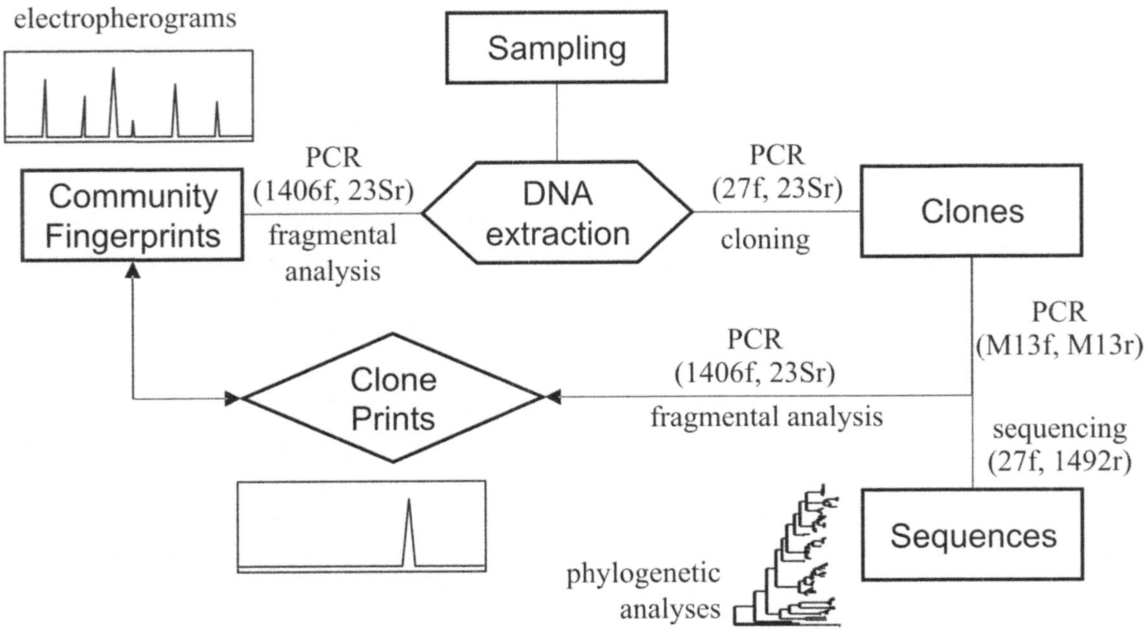

Figure 3.2 Overview of methodological approach.

The characteristics of the 16S rRNA and ITS phylogenetic markers used in this research were the base sequence of the 16S rRNA gene and the fragment length of the ITS region. The former reveals the phylogenetic identity and relatedness of detected organisms, and is commonly used in microbial ecological studies (Olsen and Woese 1993; Amann and Ludwig 2000). The latter yields community fingerprints, and was only recently introduced as a high-throughput phylogenetic marker (Fisher and Triplett 1999). To sequence the 16S rRNA gene and determine the ITS length, the targeted genes need to be amplified for easy detection. Amplification is done with the polymerase chain reaction (PCR), which employs specific forward and reverse primers targeting the two matching strands of genomic DNA at locations that flank the targeted genes (see Table 3.1 for the list of primers and Figure 3.1 for primer target positions). Amplified genes can then be processed and analyzed with respect to sequence or fragment length.

An overview of the methodological framework is provided in Figure 3.2. For the sequencing of the 16S rRNA gene, a large fragment containing nearly the whole 16S and also the ITS region was amplified using the 27f and 23Sr primer set (Figures 3.1 and 3.2). Although it was not sequenced, the ITS region was included in this process to provide a connection between 16S signatures and ARISA fingerprints (see below). The amplicon was transferred to bacterial clones as inserted into a plasmid vector. As each clone carries one type of insertion, individual 16S fragments from sampled clones can be amplified with plasmid-specific primers (M13f and M13r, see Figures 3.1 and 3.2) for subsequent sequencing. The sequencing reaction is also a polymerization-like reaction conducted on a PCR thermocycler, but only one primer is necessary. Approximately 600 to 800 nucleotide-long sequences can be accurately obtained at a time. In this study, partial sequences of 16S rRNA were obtained from the 5' end using the primer 27f, and from the 3' end using 1492r (Figure 3.1). When both are available, such partial sequences overlap towards the center of the gene, and therefore, they can be combined to produce a nearly full sequence of the 16S rRNA gene.

Table 3.1
List of PCR primers used for amplification of 16S rRNA gene and ITS

Name	Target	Position*	Sequence (5'-3')	Reference
27f	16S rRNA gene	8-27†	AGAGTTTGATCMTGGCTCAG	(Giavannoni 1991)
1492r	16S rRNA gene	1492-1510†	GGYTACCTTGTTACGACTT	(Lane 1991)
1406f	16S rRNA gene	1391-1406†	TGYACACACCGCCCGT	(Lane et al. 1985)
23Sr	23S rRNA gene	115-130†	GGGTTBCCCCATTCRG	(Fisher and Triplett 1999)
M13f	pCR®4-TOPO® vector	355-370	GTAAAACGACGGCCAG	Invitrogen‡
M13r	pCR®4-TOPO® vector	205-221	CAGGAAACAGCTATGAC	Invitrogen‡

*Position of complementary site in the target molecule.
†*E. coli* numbering.
‡Company from which the vector was bought: Invitrogen, Carlsbad, CA

The partial and full sequences form a 16S library for the distribution systems studied. The result of sequencing analyses include the classification of organisms in the library based on the comparison of their 16S sequences with public databases, and phylogenetic trees showing the relatedness of sequences with each other and with cultured or uncultured organisms obtained from other habitats. These results are presented in Chapter 4.

Community fingerprints were obtained with a technique called Automated Ribosomal Intergenic Spacer Analysis (ARISA), in which, the extracted genomic DNA was targeted with a fluorescently labeled forward primer (1406f) and an unlabeled reverse primer (23Sr), and the length of the amplified fragment containing the ITS region plus flanking portions from 16S rRNA and 23S rRNA genes was determined by capillary electrophoresis (Fisher and Triplett 1999). ARISA electropherograms that show the different sizes of fragments detected from different microorganisms constitute the microbial community fingerprint of a given sample. Comparisons of ARISA fingerprints obtained from different systems and at different times reveal variability of bacterial communities within systems and between systems. The results of such analyses can be found in Chapter 5.

In this project, a connection between 16S sequences and community fingerprints was established by matching sequenced clones with fragment peaks in ARISA profiles. Since the ITS region was included in the cloning procedure (Figure 3.1), the ITS length of a cloned organism could be determined by performing ARISA on the clone. Therefore, some of the variability observed in ARISA profiles could be linked to the absence and presence of specific organisms. This connection is detailed in Chapter 6.

SELECTION OF EXPERIMENTAL METHODS

In molecular ecology, the selection of appropriate methods is a project-dependent decision making process. In this study, sequencing was necessary since the available 16S database of heterotrophs in chloraminated drinking water was small. In fact, the extension of this database was one of the main objectives of this project.

While 16S-based identification with sequencing is a straightforward and almost standard approach in molecular ecology, it is limited by resources. Therefore, a community fingerprinting approach was necessary to complement the analysis of microbial composition. While community fingerprints are much less informative on the specific identification of organisms, they require much less resources, allowing statistically meaningful analyses. Several options are available for community fingerprinting. Perhaps, the most popular technique is denaturing gradient gel electrophoresis (DGGE), in which different organisms are represented by different DNA bands on a gel, whose positions depend on the structure of the DNA under denaturing conditions (Muyzer et al. 1993; Vallaeys et al. 1997). The target molecule is a relatively short fragment of the rRNA gene or other phylogenetic marker. DGGE has been successfully applied to the analysis of drinking water communities (Hoefel et al. 2005; Gagnon et al. 2007). Another common fingerprinting technique is Terminal Restriction Fragment Length Polymorphism (T-RFLP) (Bruce 1997; Liu et al. 1997), a method that relies on the digestion of a labeled PCR amplicon with restriction enzymes, yielding sample-specific patterns of fragment lengths. Fragment length is detected using capillary or gel electrophoresis. T-RFLP has been used for the analysis of nitrifying communities in chloraminated drinking water (Regan et al. 2002; Regan et al. 2003), and it was the original choice for community fingerprinting in this project. However, the available software for fragment identification (Kent et al. 2003) did not produce meaningful results in preliminary experiments (data not shown).

The emerging ARISA technique was then selected over the more commonly used DGGE and T-RFLP fingerprinting methods, because of our familiarity with the technique, as it was being used by others in our laboratory. Quantitative ARISA profiles can be obtained in a high throughput fashion using capillary electrophoresis.

The molecular methods used in this study, as well as others, are not free of biases (Forney et al. 2004). This is especially true for PCR-based methods, because of the well-known PCR biases (von Wintzingerode et al. 1997). PCR amplifications were inevitable in the analysis of drinking water samples due to the low abundance of cells, and consequently, that of the targeted genes. Thus, the resolution of microbial communities is limited by the ability of the primers to amplify diverse sequences. The universal primers targeting the 16S-rRNA gene are capable of capturing most organisms as can be verified by the very large databases of 16S sequences. It is the 23-rRNA targeting primers used in ARISA that may raise doubts about the coverage of fingerprints, since 23S databases used in the design of these primers are relatively small. Indeed, the use of different primer sets has been shown to have different types of coverage in the bacterial population (Cardinale et al. 2004), although the ecological patterns are said to be conserved despite these differences (Jones et al. 2007). In addition, the extraction of genomic DNA from mixed community is a step that might miss some organisms resistant to the shear forces applied during this process. Nonetheless, the coverage of molecular techniques is generally considered to be superior to any culture-based methods due to the fact that most organisms are not cultivated yet (Amann et al. 1995).

Finally it should be noted that the sequencing and fingerprinting methods do not give direct quantitative information about the abundance of different organisms in the microbial communities studied. This is also due to the biases introduced during DNA extraction and PCR-based amplification of the genes. Therefore, results should rather be interpreted from an absence/presence point of view. However, the approach can be considered as semi-quantitative in the sense that comparative analyses within- and between-distribution systems can give a firm idea about whether a detectable organism is selected for or selected against, since the methodological biases should be constant for all samples. In what follows, we provide further details of materials and methods in the experimental protocol, and give a description of data analysis techniques used.

EXPERIMENTAL PROTOCOL

DNA Extraction

For extracting the genomic DNA from drinking water samples, cells were first concentrated by filtering water through a 0.20 micron pore size nylon membrane filter (Whatman International Ltd., Maidstone, England). The filtered volume was between 300 and 500 mL for bench-scale reactors, 500 and 2000 mL for pilot-scale distribution system simulators, and 2000 mL for all full-scale samples.

Filters with concentrated cells were then stripped with a razor and processed for DNA extraction and purification using an Ultra Clean Soil DNA extraction kit (MoBio Laboratories, Solana Beach, CA). The filter-attached cells were homogenized by bead beating, and the rest of the extraction was carried out following manufacturer's instructions. The advantages of using the extraction kit included effective removal of PCR inhibitors from samples and the ability to do bead beating for optimum yield of DNA.

Cloning and Sequencing

The rRNA operon fragment used in cloning and sequencing was amplified using primers 27f and 23Sr (Figure 3.1 and Table 3.1) following a standard PCR procedure. Reaction mixtures contained 1XPCR buffer (Promega, Madison, WI), 2.5 mM $MgCl_2$, 800 μM uniform dNTP mix, 400 μM of each primer, 2.5U *Taq* polymerase, and 4 μL of template DNA (app. 350 ng/ μL) in a final volume of 50 μL. PCR reactions were carried out in a Mastercycler (Eppendorf Scientific Inc., Westburg, NY). The incubation started with an initial denaturation at 94°C for 2 min; followed by 30 cycles of amplification at 94°C for 15 s, 55°C for 15 s, and 72°C for 45 s; and ended with a final extension at 72°C for 2 min. With some full-scale samples, this protocol did not yield sufficient DNA, and therefore, an additional PCR step of 15 to 30 cycles was implemented using fresh buffer and the amplicon from the first PCR pass as template.

Total PCR product from two reactions of the same DNA extraction template (100 μL) was purified using a QIAquick Gel Extraction kit (QIAGEN Sciences, Germantown, MD). The pure amplicon was then inserted into plasmids with a TOPO TA cloning kit (Invitrogen, Carlsbad, CA) and subsequently transformed into One Shot TOP10 chemically competent *Escherichia coli* (Invitrogen). Transformed cells were grown on agar plates with Difco LB Broth (Becton Dickinson, Sparks, MD) containing 50 μg/mL Kanamycin (Fisher Scientific, Fair Lawn, NJ) and app. 70 μg/mL xgal (Apex BioResearch Products, Research Triangle Park, NC). White colonies were randomly picked for subsequent growth in solution using the same LB broth. After 10-12 hours of incubation, plasmid inserts were amplified using the M13f and M13r primers (Table 3.1) in the 30-cyle PCR protocol described above, except that the final volume was 25 μL (concentrations remained the same). The product was purified with the Ampure purification kit (Agencourt Bioscience Corporation, Beverly, MA) and used as template for a standard sequencing reaction with 27f or 1492r primers, which was followed by routine sequencing analysis carried out at University of Wisconsin Biotechnology Center. In the standard sequencing methods used, labeled nucleotides at the ends of synthesized fragments indicate nucleotide identity at the position corresponding to the length of the fragment. The results are obtained in the form of chromatogram files. Approximately 600-900 base pairs (bp)-long partial sequences could be accurately obtained at this step. The purified DNA used as template for sequencing was stored at -20°C.

ARISA Procedures

Fragments covering the ITS were amplified using the bacterial primers 1406f and 23Sr (Figure 3.1 and Table 3.1). 1406f was labeled at the 5' end with fluorescein phosphoramidite for the fluorescence-based detection of fragments. PCR reactions were carried out following the protocol described above, except that the final volume was 25 μL (concentrations remained the same).

The processing of the PCR product for fragmental analysis was as explained by Fisher and Triplett (1999) with slight modifications. Briefly, 1 μl of two times diluted (other dilution factors than two were used occasionally) product was added to a 10 μl mixture of formamide and custom size standard (Bioventures Inc., Murfreesboro, TN), and the fragments were discriminated by capillary electrophoresis using an ABI Prism 310 Genetic Analyzer (Applied Biosystems, Foster City, CA) at the University of Wisconsin Biotechnology Center.

ARISA was also applied to amplified and purified plasmid inserts from clones used for sequencing.

DATA ANALYSIS

In microbial ecological studies like this one, there is no unique path for the analysis of raw data. Because the interpretation of results is largely dependent on the path taken, it was necessary to establish a robust framework for the consistent processing of data from all drinking water samples used. In this section, we describe our data analysis framework, whose main output consists of phylogenetic trees and community fingerprints (Figure 3.2).

16S Sequences

The phylogenetic information contained in 16S sequence data was extracted at multiple steps (Figure 3.3) with the help of a public web server associated with the Ribosomal Database Project (RDP) (Cole et al. 2005; Cole et al. 2007). RDP uses a comprehensive but expanding database, which contained more than 400,000 sequences as of November 2007. In addition, it offers diverse tools for processing raw sequencing data.

Initially, electrophoretic output from the sequencer was uploaded to a specific account in the RDP server, and transferred to RDP pipeline, which is a specific tool for processing electropherogram files from sequencing analyses. The PHRED algorithm (Ewing and Green 1998; Ewing et al. 1998) was used for calling the bases, and the resulting raw sequence data was polished with the LUCY algorithm (Chou and Holmes 2001), which trims end sections with low confidence in base calling. Sequences that received low PHRED quality scores were rejected.

In addition, sequences lower than 500 bp after trimming were rejected, and the sequences passing this initial processing and quality checking step were passed downstream (Figure 3.3).

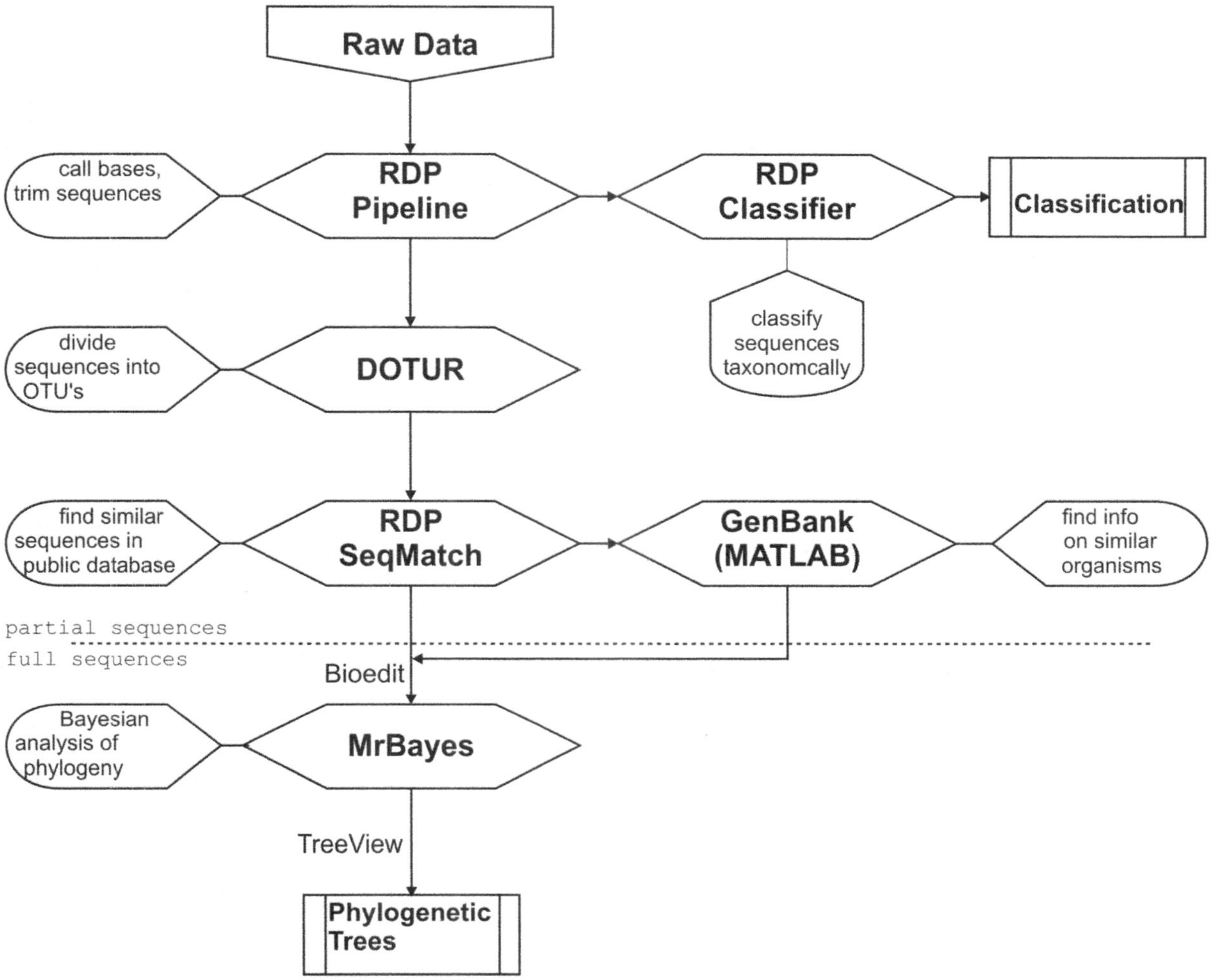

Figure 3.3 Flow chart describing the analysis of 16S rRNA sequences

Accepted sequences were next aligned with respect to other public sequences in RDP, using the RDP aligner. The aligning algorithm in RDP takes into account the secondary structure of 16S rRNA, as estimated based on reference organisms. This prevents loss of valuable phylogenetic information not directly available in primary structures (i.e. the sequences). The aligned sequences were taxonomically classified using the RDP classifier tool. The classification determines the identity of organisms detected in drinking water samples at different levels of taxa such as phylum (e.g. *Proteobacteria*), class (e.g. Gammaproteobacteria), and genus (e.g. *Escherichia*). RDP classifier gives a confidence level for the classification at each level. While these classifications form a useful output for this project, sequences were further processed for more detailed phylogenetic analysis aiming at the construction of phylogenetic trees (Figure 3.3).

To simplify the phylogenetic analysis, aligned partial sequences were clustered into operational taxonomic units (OTUs), based on Jukes-Cantor-corrected distance matrices, using the DOTUR program (Schloss and Handelsman 2005). Since the sequences were partial only, a conservative similarity cutoff value of 99% (i.e. 1% distance in the sequence distance matrix), was used to assign sequences to the same OTU.

To find public sequences that closely matched our clones, partial sequences representing each OTU were entered to the SeqMatch tool of RDP. Genbank (Benson et al. 2006) accession codes of matching sequences were obtained from the output of SeqMatch. Subsequently, these codes were used as an input to a MATLAB (The MathWorks Inc., Natick, MA) script, which searched the Genbank (http://www.ncbi.nlm.nih.gov/Genbank/index.html) and Pubmed (http://www.ncbi.nlm.nih.gov/sites/entrez?db=PubMed) databases for collecting information about the organisms (name, taxonomy, environment) and their publication if available (reference, title, abstract). Organisms that were found to be relevant, such as those retrieved from other drinking water environments, or those that were almost identical (ca. 100% sequence similarity) to OTUs in our library (irrespective of their environment) were selected to be used in the steps to follow.

While all operations up to this point were carried out using partial sequences from the 5' end of 16S rRNA, subsequent steps were taken with nearly full sequences of selected OTUs and closely matching organisms to ensure reliable clustering in the resulting phylogenetic trees. Only representatives of OTUs (i.e., one sequence per OTU) were fully sequenced and used in trees. In addition, not all OTUs were considered. Selection of OTUs to be clustered depended on the number of sequences in the OTU, the environment from which matching sequences were retrieved, etc. (the details of selection criteria will be presented when necessary, in Chapter 4). Thus, representatives of selected OTUs from our clone library and relevant organisms from public databases were gathered to form a unified sequence alignment file. Using Bioedit (Hall 1999) as supporting software, flanking ends of the alignment were trimmed to concentrate the analysis to the same range on 16S rRNA positioning for all sequences. An output of NEXUS format was exported from Bioedit as an input to MrBayes 3.2.1. (Ronquist and Huelsenbeck 2003) for the Bayesian inference of phylogeny using the General Time Reversible evolution model. The analysis was constantly run with four independent chains and for at least 400,000 generations to reach a convergence, of which, about the first 30% were discarded using the "burn-in" command. Finally, Treeview (http://taxonomy.zoology.gla.ac.uk/rod/treeview.html) was used to generate visual phylogenetic trees from MrBayes output (Figure 3.3).

All sequences generated in this study were deposited in GenBank with accession codes EU808049-EU809416. The given names of sequences included abbreviated distribution system name followed by the clone ID (date-number) and the OTU (number). For example Pro_CL-090546_OTU-16 designates the clone number 46 among those retrieved from Prospect tank in September 2005, which was assigned to OTU number 16 among all Prospect OTUs. Nearly full sequences are indicated by a "group" name ("nearly full sequence") under the FEATURES section of genbank files, which otherwise reads "partial sequence".

ARISA Profiles

A flow chart of how ARISA data was processed is given in Figure 3.4. The main idea was to convert data into observed fragment sizes in a given sample, which defines the community fingerprint of the sample, and statistically and visually evaluate similarities and differences between different samples based on these fingerprints.

Raw ARISA data consists of fluorescence intensity peaks collected at different positions during a capillary electrophoresis run. The peaks are to be positionally aligned with a size standard for determining what fragment size they correspond to. The alignment and size calling were both managed with a software named GeneMarker (SoftGenetics, State College, PA). Figure 3.5 illustrates an example ARISA electropherogram after alignment and size calling (size standard not shown). In GeneMarker, a panel common to all drinking water samples was created using a superimposition of aligned profiles for the binning of the fluorescence peaks. Those peaks with a height less than a threshold set based on estimations of noise and background with negative controls (i.e. samples without DNA template) were rejected unless they could be clearly called. Information on areas and distribution of the binned peaks was exported from GeneMarker to M.S. Excel for further data analysis (Figure 3.4). In Excel, individual sample data was normalized with respect to greatest peak height (i.e., highest peak is associated with 1), total peak area (i.e. total area becomes 1), or reduced to absence/presence information (i.e. 1 if peak height greater than threshold, and 0 otherwise). The first two normalization methods assume ARISA is quantitative although PCR-based methods are generally biased due to exponential amplification. Comparative analyses suggested that absence/presence data provided better statistical separation between different systems despite the removal of possibly false quantitative information. Thus, only results with absence/presence normalization are reported in Chapter 5, where ARISA analysis is presented.

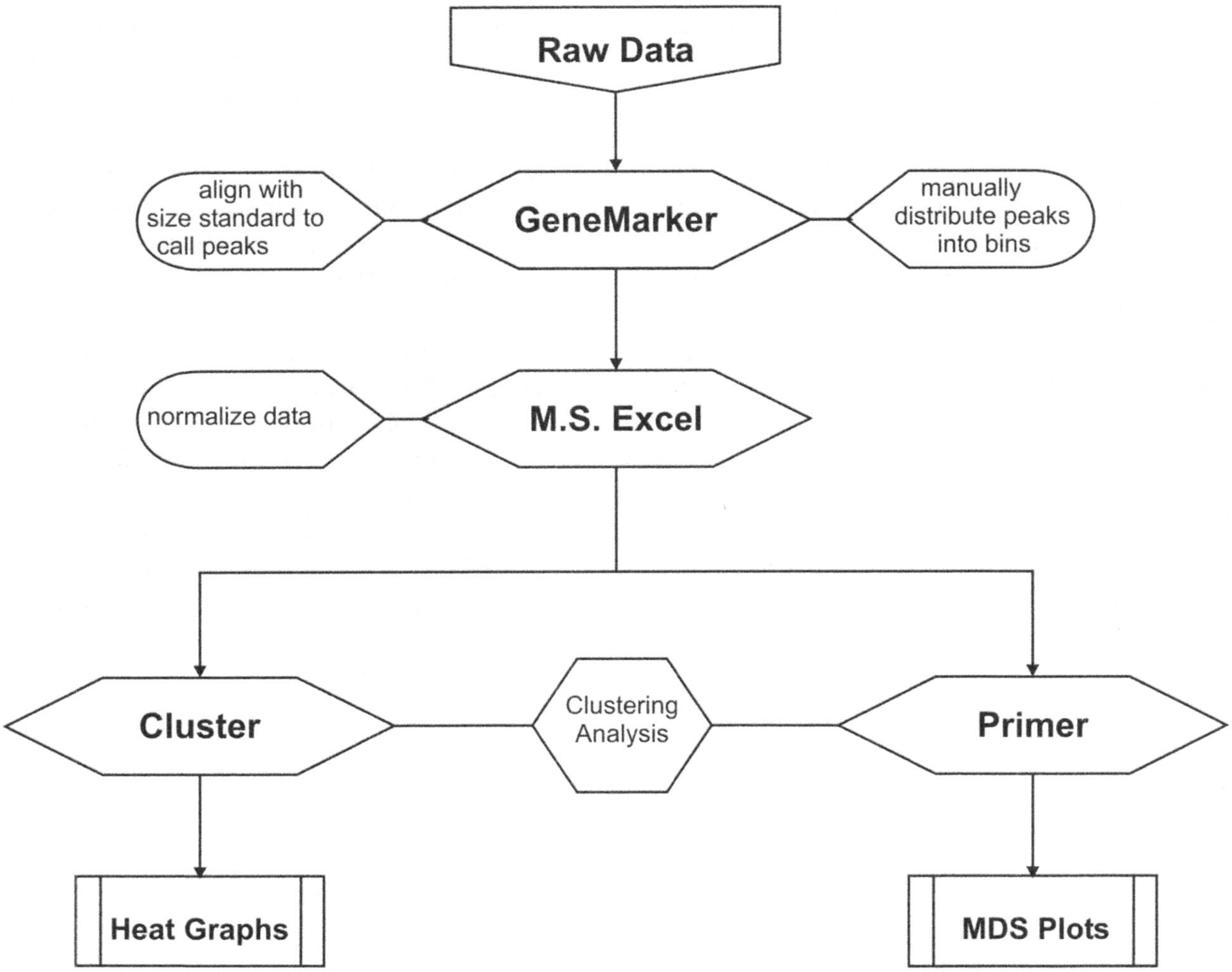

Figure 3.4 Flow chart describing the analysis of ARISA profiles

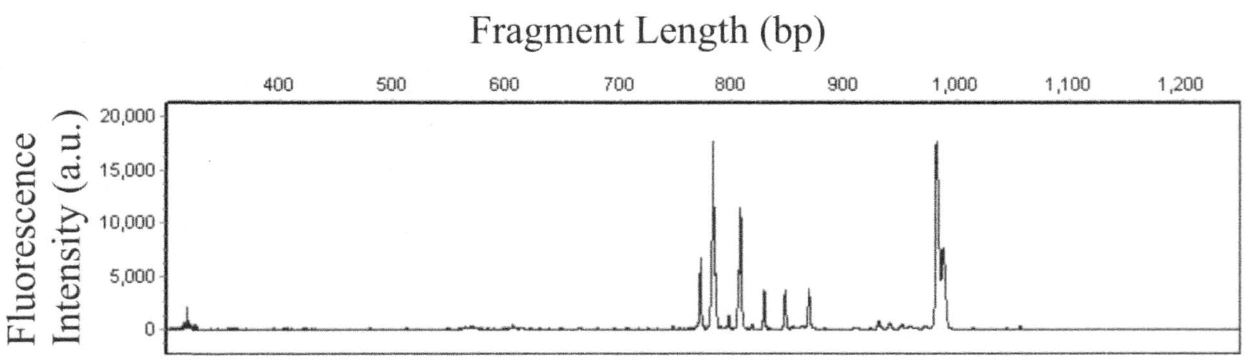

Figure 3.5 Example ARISA electropherogram

The normalized data from a set of samples were clustered using two different types of software: Cluster 3.0 (Eisen et al. 1998; de Hoon et al. 2004) and Primer 5.2.7 (Primer-E Ltd., Roborough, Plymouth, United Kingdom). With Cluster 3.0, hierarchical clustering methods were practiced to create dendrograms in a heat map representation of normalized data (see

Chapter 5 for the images). The similarity of profiles was quantified based on uncentered correlation, and dendrograms were formed using the average-linkage algorithm. With Primer 5, a non-metric multidimensional scaling (MDS) method was used as described elsewhere (Yannarell and Triplett 2004). First, an n-by-n triangular matrix (n=number of ARISA profiles) defining the similarity of ARISA sets was developed based on Bray-Curtis similarity of Lengendre and Legendre (Legendre and Legendre 1998). Then, a 2D projection of the multi-dimensional clustering was created (see Chapter 5 for the image). The compromise from actual sample-sample distances during the projection was evaluated based on Kruskal stress formula (Kruskal 1964), which indicates smaller risk of misinterpretation as values get closer to 0.

In addition to what is described above for data analysis, diversity and species richness of samples based on 16S sequences and ARISA profiles were evaluated. The details of such analyses will be provided when relevant, in the chapters to follow. Finally, it should be stated that other tools and methodological approaches were evaluated throughout the project before deciding on the final analyses and algorithms illustrated in Figures 3.3 and 3.4. These trial-and-error steps are not described in the report.

CHAPTER 4
RESULTS AND DISCUSSION: PHYLOGENETIC ANALYSES BASED ON RIBOSOMAL RNA SEQUENCES

The first level of analysis in this project was based on the comparison of 16S rRNA sequences retrieved from the different chloraminated environments studied, which included two bench-scale reactors, four pilot-scale distribution systems, and three locations in a full scale distribution system. Figure 4.1 presents a schedule of sampling for this task. Sampling for sequencing took place between April 2005 and April 2007.

The library of 16S sequences generated was analyzed using three approaches:

1. For each tested environment, sequences were arranged according to their taxonomic hierarchy. This was useful to identify what phylogenetic groups were present and their frequency within the library.
2. Sequences in the library were compared to publicly available databases of RDP (Cole et al. 2005) and GenBank (Benson et al. 2006) to identify instances when closely matching organisms have been observed before. Through this level of analysis, we extracted from the databases information related to the environments where the closely matching organisms were identified.
3. The library of sequences was organized into OTUs according to their similarity to each other. Representatives of each OTU were then selected for the construction of phylogenetic trees that allowed the visualization of relatedness of the defined OTUs to each other and to well-defined phylogenetic groups as well as to uncultured bacteria.

Scale	Name	2005									2006												2007			
		4	5	6	7	8	9	10	11	12	1	2	3	4	5	6	7	8	9	10	11	12	1	2	3	4
Bench	Chl-						■						■													
	Chl+	■					■			■			■													
Pilot	UWL					■			■										■		■		■			
	UWH					■			■									■		■			■			
	LO1																			■		■				
	LO3																			■						■
Full	Finished														■					■					■	
	Martin						■		■						■					■					■	
	Prospect						■		■						■					■					■	

Figure 4.1 Sampling schedule for sequencing

While there are significant overlaps in the information obtained with these different analyses, each type of analysis contributes to the understanding of the diversity of bacterial populations in the different environments studied. The goal of this chapter is to present the information retrieved from the sequence analysis in an organized manner, without attempting to perform an in depth analysis of the meaning of each identified sequence.

TAXONOMIC CLASSIFICATION BASED ON PARTIAL SEQUENCES

The classification of partial sequences retrieved from the clone library is presented in Tables 4.1 to 4.9. Each table summarizes the sequences retrieved from a specific environment studied. The assignments are based on a 95% confidence level (Cole et al. 2005). The number of total sequences assigned to taxonomic groups decreases in going from higher to lower levels of hierarchy, as some sequences cannot be classified with 95% confidence in the next level. Thus, some sequences with low quality of nucleotide identification get eliminated at higher divisions such as phylum and class. However, unclassified sequences at lower levels do not necessarily indicate poor quality of the sequence, but may rather point to the presence of sequences that represent microorganisms not yet classified in the RDP database. As one of the measures of quality, sequence datasets were tested in Bellerophon (Huber et al. 2004) for detection of potential chimeras, and no relationship was observed between unclassified sequences and chimera scores. A total of only 3 sequences were considered as potential chimeras and removed from the clone library.

Some of the 16S rRNA sequences in the library were virtually identical (Similarity = 1.000) to published sequences in the RDP database. For these sequences, information was retrieved from their corresponding GenBank entries, including the environmental or health relevance of the matching organism and the title of the reference describing the public sequence. This information is presented in Table 4.10. Out of 55 perfect matching sequences described in Table 4.10, six of them were initially described as being related to a drinking water environment, 8 were related to either fresh water or groundwater environments, 3 corresponded to observations in marine environments, 2 were related to activated sludge environments, and 3 were similar to those found in a clinical environment.

The sequence classifications shown in Tables 4.1 to 4.9 are useful for observing the similarities and differences in different systems with respect to the taxonomic divisions present. They are also useful to assess which taxonomic groups are not represented in the library.

At the phylum level, all environments tested included the presence of sequences representing the phyla *Actinobacteria*, *Bacteroidetes*, and *Proteobacteria*, and in general, these three phyla represented the majority of the sequences in the library. Sequences belonging to the phyla *Nitrospira* and *Firmicutes* were detected in 5 out of the 9 environments tested. *Spirochaetes* were identified in 2 environments, while *Acidobacteria* were detected in only one environment. It is interesting to note that in the most recent release of the RDP database (June, 2008) there are 34 recognized phyla within the domain Bacteria, while the library generated in this project contained sequences from only 7 phyla.

Among *Actinobacteria*, sequences belonging to the genus *Mycobacterium* were found in all systems, with remarkable frequency in all the pilot-scale distribution system samples and in the samples from the Martin's Hill Tank. The members of this genus have significant relevance to drinking water treatment (Berry et al. 2006), as they are notorious for disinfection resistance (Carson et al. 1978; Taylor et al. 2000), are found in biofilm samples (Dailloux et al. 2003), and some are known pathogens (Falkinham et al. 2001). Furthermore, they were previously reported to be found in drinking water samples (Falkinham et al. 2001; Dailloux et al. 2003; Williams et al. 2004) and possibly with increasing frequency in chloraminated environments (Pryor et al. 2004).

Most sequences under the phylum *Bacteroidetes* were assigned to the *Flavobacteriales* and *Sphingobacteriales* families, and a lower number of sequences to the *Bacteroidales* family.

It is interesting to note that all sequences related to the *Bacteroidales* family belonged to the *Dysgonomonas* genus, and that they were only detected in samples received from Louisville (pilot-scale and full-scale samples), so their presence may have a particular geographical meaning or could be representative of the source of water. Also noteworthy is that a large group of unclassified *Sphingobacteriales* was found in the chloraminated Chl+ chemostat (Table 4.2). Most of these were obtained in the last sample from this reactor, where the chloramine concentration reached the highest level (see Chapter 2). Interestingly, Gagnon et al. (2007) reported the presence of *Cytophaga sp.*, an organism under the same family, in an annular bench-scale reactor with high chloramine dose, but not in untreated or low-dose reactors. Thus, some members of the *Sphingobacteriales* family might be resistant to chloramination.

Sequences under the phylum *Firmicutes* were particularly prevalent in samples from the full-scale system, with *Bacilli* and *Mollicutes* being the representative classes. Among the *Bacilli* class, members of the genus *Staphylococcus*, *Dolosigranulum*, and *Streptococcus* were recurrent. All sequences within the *Mollicutes* were classified in the *Erysipelotrichaceae* family, but were not classified in any specific genus.

Sequences assigned to the *Nitrospira* phylum were found in bench-scale, pilot-scale, and full-scale samples, but were a minor component of the library. In all cases, clones were classified under the *Nitrospira* genus, which includes nitrite oxidizing bacteria (Regan et al. 2002; Regan et al. 2003), and therefore are indicators of nitrification in these systems.

By far, the majority of the sequences that could be assigned to a phylum belonged to *Proteobacteria.* Sequences were generally distributed among the *Alpha-*, *Beta-*, and *Gammaproteobacteria*, indicating a large share of these classes consistent with other clone libraries of drinking water (Williams et al. 2004; Tokajian et al. 2005), although *Deltaproteobacteria* was also detected in the pilot scale system operated at the University of Wisconsin and in one of the full-scale samples. Among the *Alphaproteobacteria*, members of the *Caulobacterales*, *Rhizobiales*, and *Sphingomonadales* appeared to be the significant orders in most samples. *Betaproteobacteria* included clones belonging to the *Nitrosomonas* genus associated with ammonia oxidation (Regan et al. 2002). Most of the other *Betaproteobacteria* were classified under the *Burkholderiales* order, with the *Limnobacter* genus being a frequent observation. The presence of *Gammaproteobacteria* was also common. A large number of *Stenotrophomonas*-related clones in the bench-scale reactors is believed to be the result of contamination of the reactors with a pure culture of *Stenotrophomonas maltophilia* that was being used in the laboratory at the time of sampling. Another large bench-scale group of *Gammaproteobacteria* clones was formed by *Acinetobacter*-related sequences, but were observed only in the Chl- reactor, possibly indicating the sensitivity of these organisms to chloramination. However, Gagnon et al. (2007) reported the presence of *Acinetobacter*-related species in bench-scale annular reacors with low and high doses of chloramines, as well as those without treatment. Finally, the *Legionella*- and *Pseudomonas*- related sequences are noteworthy due to the presence of some pathogenic species under these genera (WHO 2003; Berry et al. 2006).

Table 4.1
Classification of identified partial sequences[*] for Bacteria[†] obtained from the non-chloraminated bench-scale chemostat Chl-

Phylum	Class	Order	Family	Genus
Actinobacteria (6)	*Actinobacteria* (6)	*Actinomycetales* (6)	*Micrococcaceae* (3)	*Arthrobacter* (3)
			Mycobacteriaceae (2)	*Mycobacterium* (2)
			Nocardiaceae (1)	*Rhodococcus* (1)
Bacteroidetes (12)	*Flavobacteria* (7)	*Flavobacteriales* (7)	*Flavobacteriaceae* (7)	*Chryseobacterium* (1)
				Flavobacterium (5)
	Sphingobacteria (5)	*Sphingobacteriales* (5)	*Flexibacteraceae* (1)	*Emticicia* (1)
			Sphingobacteriaceae (3)	*Pedobacter* (1)
				Sphingobacterium (2)
Firmicutes (1)	*Mollicutes* (1)	*Incertae_sedis_8* (1)	*Erysipelotrichaceae* (1)	-
Nitrospira (2)	*Nitrospira* (2)	*Nitrospirales* (2)	*Nitrospiraceae* (2)	*Nitrospira* (2)
Proteobacteria (64)	*Alphaproteobacteria* (6)	*Caulobacterales* (1)	*Caulobacteraceae* (1)	*Brevundimonas* (1)
		Rhizobiales (4)	*Bradyrhizobiaceae* (3)	*Bradyrhizobium* (2)
			Hyphomicrobiaceae (1)	*Devosia* (1)
	Betaproteobacteria (12)	*Burkholderiales* (12)	*Alcaligenaceae* (1)	*Pigmentiphaga* (1)
			Burkholderiaceae (7)	*Limnobacter* (7)
			Comamonadaceae (3)	*Acidovorax* (3)
	Gammaproteobacteria (46)	*Pseudomonadales* (39)	*Moraxellaceae* (38)	*Acinetobacter* (38)
			Pseudomonadaceae (1)	*Pseudomonas* (1)
		Xanthomonadales (7)	*Xanthomonadaceae* (7)	*Pseudoxanthomonas* (1)
				Stenotrophomonas (6)

*For each taxa, the number of sequences identified at a 95% confidence level is shown in parenthesis.
†A total of 85 Bacteria sequences were obtained from chemostat Chl-. All sequences were classified at the phylum and class levels, 84 sequences at the order level, 82 at the family level, and 79 at the genus level.

Table 4.2

Classification of identified partial sequences* for Bacteria† obtained from the chloraminated bench-scale chemostat Chl+

Phylum	Class	Order	Family	Genus
Actinobacteria (39)	*Actinobacteria* (39)	*Actinomycetales* (39)	*Gordoniaceae* (4)	*Gordonia* (4)
			Microbacteriaceae (3)	*Microbacterium* (3)
			Micrococcaceae (3)	*Arthrobacter* (3)
			Mycobacteriaceae (25)	*Mycobacterium* (25)
			Nocardiaceae (4)	*Rhodococcus* (4)
Bacteroidetes (53)	*Flavobacteria* (7)	*Flavobacteriales* (7)	*Flavobacteriaceae* (7)	*Chryseobacterium* (6)
	Sphingobacteria (46)	*Sphingobacteriales* (46)	*Flexibacteraceae* (2)	*Emticicia* (1)
				Spirosoma (1)
Nitrospira (2)	*Nitrospira* (2)	*Nitrospirales* (2)	*Nitrospiraceae* (2)	*Nitrospira* (2)
Proteobacteria (79)	*Alphaproteobacteria* (50)	*Caulobacterales* (19)	*Caulobacteraceae* (19)	*Brevundimonas* (2)
				Caulobacter (17)
		Rhizobiales (5)	*Bradyrhizobiaceae* (4)	*Afipia* (2)
				Bosea (2)
			Hyphomicrobiaceae (1)	*Hyphomicrobium* (1)
		Sphingomonadales (21)	*Sphingomonadaceae* (21)	*Novosphingobium* (3)
				Sphingomonas (2)
				Sphingopyxis (15)
	Betaproteobacteria (16)	*Burkholderiales* (12)	*Alcaligenaceae* (9)	*Achromobacter* (9)
			Comamonadaceae (3)	*Acidovorax* (1)
				Variovorax (2)
		Methylophilales (1)	*Methylophilaceae* (1)	*Methylophilus* (1)
		Nitrosomonadales (3)	*Nitrosomonadaceae* (3)	*Nitrosomonas* (3)
	Gammaproteobacteria (13)	*Legionellales* (1)	*Coxiellaceae* (1)	*Coxiella* (1)
		Pseudomonadales (2)	*Pseudomonadaceae* (2)	*Pseudomonas* (2)
		Xanthomonadales (10)	*Xanthomonadaceae* (10)	*Stenotrophomonas* (7)

*For each taxa, the number of sequences identified at a 95% confidence level is shown in parenthesis.

†A total of 174 Bacteria sequences were obtained from chemostat Chl+; 173 sequences were classified at the phylum and class levels, 168 at the order level, 124 at the family level, and 119 at the genus level.

Table 4.3

Classification of identified partial sequences* for Bacteria† obtained from pilot-scale distribution system UWL

Phylum	Class	Order	Family	Genus
Actinobacteria (29)	*Actinobacteria* (29)	*Actinomycetales* (29)	*Mycobacteriaceae* (27)	*Mycobacterium* (27)
Bacteroidetes (18)	*Flavobacteria* (4)	*Flavobacteriales* (4)	*Flavobacteriaceae* (4)	*Flavobacterium* (4)
	Sphingobacteria (6)	*Sphingobacteriales* (6)	-	-
Nitrospira (2)	*Nitrospira* (2)	*Nitrospirales* (2)	*Nitrospiraceae* (2)	*Nitrospira* (2)
Proteobacteria (158)	*Alphaproteobacteria* (130)	*Caulobacterales* (5)	*Caulobacteraceae* (5)	*Phenylobacterium* (5)
		Rhizobiales (9)	*Bradyrhizobiaceae* (2)	-
			Hyphomicrobiaceae (2)	*Hyphomicrobium* (1)
		Rickettsiales (1)	*Incertae_sedis_4* (1)	*Caedibacter* (1)
		Sphingomonadales (92)	*Sphingomonadaceae* (92)	*Novosphingobium* (3)
				Porphyrobacter (2)
				Sphingomonas (13)
				Sphingopyxis (5)
	Betaproteobacteria (21)	*Burkholderiales* (16)	*Burkholderiaceae* (16)	*Limnobacter* (16)
		Nitrosomonadales (3)	*Nitrosomonadaceae* (3)	*Nitrosomonas* (3)
		Rhodocyclales (1)	*Rhodocyclaceae* (1)	-
	Deltaproteobacteria (2)	*Bdellovibrionales* (2)	*Bdellovibrionaceae* (2)	*Bdellovibrio* (2)
	Gammaproteobacteria (5)	*Legionellales* (2)	*Legionellaceae* (2)	*Legionella* (2)
		Xanthomonadales (2)	*Xanthomonadaceae* (2)	-

*For each taxa, the number of sequences identified at a 95% confidence level is shown in parenthesis.

†A total of 221 Bacteria sequences were obtained from UWL; 207 were classified at the phylum level, 199 at the class level, 174 at the order level, 161 at the family level, and 86 at the genus level.

Table 4.4

Classification of identified partial sequences[*] for Bacteria[†] obtained from pilot-scale distribution system UWH

Phylum	Class	Order	Family	Genus
Acidobacteria (2)	*Acidobacteria* (2)	*Acidobacteriales* (2)	*Acidobacteriaceae* (2)	Gp3 (2)
Actinobacteria (44)	*Actinobacteria* (44)	*Actinomycetales* (44)	*Mycobacteriaceae* (44)	*Mycobacterium* (44)
Bacteroidetes (6)	*Flavobacteria* (1)	*Flavobacteriales* (1)	*Flavobacteriaceae* (1)	*Chryseobacterium* (1)
	Sphingobacteria (2)	*Sphingobacteriales* (2)		
Proteobacteria (158)	*Alphaproteobacteria* (79)	*Caulobacterales* (5)	*Caulobacteraceae* (5)	*Brevundimonas* (1)
				Caulobacter (1)
				Phenylobacterium (3)
		Rhizobiales (15)	*Bradyrhizobiaceae* (1)	-
			Hyphomicrobiaceae (12)	*Hyphomicrobium* (4)
		Rhodospirillales (1)	*Acetobacteraceae* (1)	*Roseomonas* (1)
		Sphingomonadales (35)	*Sphingomonadaceae* (35)	*Blastomonas* (2)
				Novosphingobium (4)
				Porphyrobacter (7)
				Sphingomonas (14)
	Betaproteobacteria (18)	*Burkholderiales* (10)	*Burkholderiaceae* (9)	*Limnobacter* (9)
			Incertae_sedis_5 (1)	-
		Methylophilales (4)	*Methylophilaceae* (4)	*Methylophilus* (4)
		Nitrosomonadales (4)	*Nitrosomonadaceae* (4)	*Nitrosomonas* (4)
	Deltaproteobacteria (2)	*Bdellovibrionales* (2)	*Bdellovibrionaceae* (2)	*Bdellovibrio* (2)
	Gammaproteobacteria (56)	*Chromatiales* (1)		
		Legionellales (1)	*Legionellaceae* (1)	*Legionella* (1)
		Pseudomonadales (24)	*Pseudomonadaceae* (24)	*Pseudomonas* (24)

*For each taxa, the number of sequences identified at a 95% confidence level is shown in parenthesis.

†A total of 215 Bacteria sequences were obtained from UWH; 210 were classified at the phylum level, 204 at the class level, 151 at the order level, 146 at the family level, and 128 at the genus level.

Table 4.5

Classification of identified partial sequences[*] for Bacteria[†] obtained from Louisville's pilot-scale distribution system LO1

Phylum	Class	Order	Family	Genus
Actinobacteria (58)	*Actinobacteria* (58)	*Actinomycetales* (58)	*Mycobacteriaceae* (58)	*Mycobacterium* (58)
Bacteroidetes (3)	*Sphingobacteria* (1)	*Sphingobacteriales* (1)	*Flexibacteraceae* (1)	-
Proteobacteria (33)	*Alphaproteobacteria* (33)	*Caulobacterales* (2)	*Caulobacteraceae* (2)	*Caulobacter* (2)
		Rhizobiales (15)	*Hyphomicrobiaceae* (11) *Methylobacteriaceae* (1)	*Hyphomicrobium* (11) *Methylobacterium* (1)
		Sphingomonadales (7)	*Sphingomonadaceae* (7)	*Porphyrobacter* (1) *Sphingomonas* (2)

*For each taxa, the number of sequences identified at a 95% confidence level is shown in parenthesis.

†A total of 94 Bacteria sequences were obtained from UWH; 94 were classified at the phylum level, 92 at the class level, 83 at the order level, 80 at the family level, and 75 at the genus level.

Table 4.6

Classification of identified partial sequences[*] for Bacteria[†] obtained from Louisville's pilot-scale distribution system LO3

Phylum	Class	Order	Family	Genus
Actinobacteria (66)	*Actinobacteria* (66)	*Actinomycetales* (66)	*Mycobacteriaceae (66)*	*Mycobacterium* (66)
Bacteroidetes (1)	*Bacteroidetes* (1)	*Bacteroidales* (1)	*Porphyromonadaceae* (1)	*Dysgonomonas* (1)
Firmicutes (1)	*Clostridia* (1)	*Clostridiales* (1)	*Peptostreptococcaceae* (1)	*Fusibacter* (1)
Nitrospira (1)	*Nitrospira* (1)	*Nitrospirales* (1)	*Nitrospiraceae* (1)	*Nitrospira* (1)
Proteobacteria (21)	*Alphaproteobacteria* (21)	*Caulobacterales* (1)	*Caulobacteraceae* (1)	*Brevundimonas* (1)
		Rhizobiales (15)	*Hyphomicrobiaceae* (10)	*Hyphomicrobium* (10)
		Sphingomonadales (1)	*Sphingomonadaceae* (1)	*Novosphingobium* (1)

*For each taxa, the number of sequences identified at a 95% confidence level is shown in parenthesis.

†A total of 90 Bacteria sequences were obtained from UWH; 90 were classified at the phylum and class levels, 86 at the order level, and 81 at the family and genus levels.

Table 4.7
Classification of identified partial sequences* for Bacteria† obtained from finished water at Louisville

Phylum	Class	Order	Family	Genus
Actinobacteria (11)	*Actinobacteria* (11)	*Actinomycetales* (8)	*Corynebacteriaceae* (3)	*Corynebacterium* (3)
			Mycobacteriaceae (2)	*Mycobacterium* (2)
			Propionibacteriaceae (2)	*Propionibacterium* (2)
Bacteroidetes (21)	*Bacteroidetes* (4)	*Bacteroidales* (4)	*Porphyromonadaceae* (4)	*Dysgonomonas* (4)
	Flavobacteria (7)	*Flavobacteriales* (7)	*Flavobacteriaceae* (7)	*Flavobacterium* (3)
	Sphingobacteria (3)	*Sphingobacteriales* (3)	*Flexibacteraceae* (2)	*Arcicella* (2)
			Sphingobacteriaceae (1)	*Sphingobacterium* (1)
Firmicutes (27)	*Bacilli* (11)	*Bacillales* (6)	*Paenibacillaceae* (1)	*Cohnella* (1)
			Staphylococcaceae (5)	*Staphylococcus* (5)
		Lactobacillales (5)	*Aerococcaceae* (1)	*Aerococcus* (1)
			Carnobacteriaceae (2)	*Dolosigranulum* (1)
			Leuconostocaceae (1)	*Weissella* (1)
			Streptococcaceae (1)	*Streptococcus* (1)
	Clostridia (4)	*Clostridiales* (4)	*Peptococcaceae* (2)	*Desulfosporosinus* (2)
			Peptostreptococcaceae (1)	*Fusibacter* (1)
	Mollicutes (12)	*Incertae_sedis_8* (12)	*Erysipelotrichaceae* (12)	-
Proteobacteria (70)	*Alphaproteobacteria* (8)	*Caulobacterales* (1)	*Caulobacteraceae* (1)	*Caulobacter* (1)
		Rhizobiales (5)	*Brucellaceae* (3)	*Pseudochrobactrum* (2)
			Hyphomicrobiaceae (2)	*Hyphomicrobium* (2)
		Rhodospirillales (1)	*Acetobacteraceae* (1)	-
	Betaproteobacteria (21)	*Burkholderiales* (3)	*Alcaligenaceae* (1)	-
			Comamonadaceae (1)	*Rhodoferax* (1)
			Oxalobacteraceae (1)	-
		Methylophilales (1)	*Methylophilaceae* (1)	*Methylophilus* (1)
		Rhodocyclales (17)	*Rhodocyclaceae* (17)	*Azonexus* (16)

(continued)

Table 4.7 (Continued)

Phylum	Class	Order	Family	Genus
	Gammaproteobacteria (41)	*Chromatiales* (1)	*Chromatiaceae* (1)	*Rheinheimera* (1)
		Enterobacteriales (7)	*Enterobacteriaceae* (7)	-
		Oceanospirillales (8)	*Alcanivoraceae* (8)	*Alcanivorax* (8)
		Pseudomonadales (12)	*Pseudomonadaceae* (12)	*Pseudomonas* (12)
		Xanthomonadales (3)	*Xanthomonadaceae* (3)	*Silanimonas* (1)
				Stenotrophomonas (2)
Spirochaetes (2)	*Spirochaetes* (2)	*Spirochaetales* (2)	*Spirochaetaceae* (2)	*Spirochaeta* (2)

*For each taxa, the number of sequences identified at a 95% confidence level is shown in parenthesis.

†A total of 132 Bacteria sequences were obtained from the finished water at Louisville Water Company; 131 were classified at the phylum level, 124 at the class level, 110 at the order level, 108 at the family level, and 79 at the genus level.

Table 4.8
Classification of identified partial sequences* for Bacteria† obtained from the Prospect Tank at Louisville

Phylum	Class	Order	Family	Genus
Actinobacteria (10)	*Actinobacteria* (10)	*Actinomycetales* (10)	*Mycobacteriaceae* (9)	*Mycobacterium* (9)
Bacteroidetes (9)	*Bacteroidetes* (9)	*Bacteroidales* (9)	*Porphyromonadaceae* (9)	*Dysgonomonas* (9)
Firmicutes (8)	*Bacilli* (6)	*Bacillales* (3)	*Paenibacillaceae* (1)	*Brevibacillus* (1)
		Lactobacillales (3)	*Carnobacteriaceae* (3)	*Dolosigranulum* (3)
	Mollicutes (2)	*Incertae_sedis_8* (2)	*Erysipelotrichaceae* (2)	-
Nitrospira (3)	*Nitrospira* (3)	*Nitrospirales* (3)	*Nitrospiraceae* (3)	*Nitrospira* (3)
Proteobacteria (144)	*Alphaproteobacteria* (69)	*Rhizobiales* (28)	*Bradyrhizobiaceae* (2)	*Bradyrhizobium* (1)
			Methylobacteriaceae (9)	*Methylobacterium* (9)
		Sphingomonadales (19)	*Sphingomonadaceae* (19)	*Blastomonas* (1)
				Sphingomonas (9)
				Sphingopyxis (1)
	Betaproteobacteria (58)	*Burkholderiales* (15)	*Comamonadaceae* (12)	*Curvibacter* (7)
			Incertae_sedis_5 (1)	*Rubrivivax* (1)
			Oxalobacteraceae (2)	*Janthinobacterium* (1)
		Nitrosomonadales (29)	*Nitrosomonadaceae* (29)	*Nitrosomonas* (29)
		Rhodocyclales (11)	*Rhodocyclaceae* (11)	*Azonexus* (11)
	Gammaproteobacteria (16)	*Enterobacteriales* (1)	*Enterobacteriaceae* (1)	-
		Legionellales (5)	*Legionellaceae* (5)	-
		Methylococcales (6)	*Methylococcaceae* (6)	*Methylomonas* (2)
		Pseudomonadales (2)	*Pseudomonadaceae* (2)	*Pseudomonas* (2)

*For each taxa, the number of sequences identified at a 95% confidence level is shown in parenthesis.
†A total of 176 Bacteria sequences were obtained from the Prospect Tank at Louisville; 174 were classified at the phylum level, 173 at the class level, 146 at the order level, 126 at the family level, and 99 at the genus level.

Table 4.9
Classification of identified partial sequences* for Bacteria† obtained from the Martin's Hill Tank at Louisville

Phylum	Class	Order	Family	Genus
Actinobacteria (58)	*Actinobacteria* (58)	*Actinomycetales* (57)	*Corynebacteriaceae* (1)	*Corynebacterium* (1)
			Mycobacteriaceae (52)	*Mycobacterium* (52)
			Nocardiaceae (1)	*Rhodococcus* (1)
			Propionibacteriaceae (1)	*Propionibacterium* (1)
Bacteroidetes (22)	*Bacteroidetes* (1)	*Bacteroidales* (1)	*Porphyromonadaceae* (1)	*Dysgonomonas* (1)
	Flavobacteria (3)	*Flavobacteriales* (3)	*Flavobacteriaceae* (3)	*Flavobacterium* (1)
	Sphingobacteria (2)	*Sphingobacteriales* (2)	*Flexibacteraceae* (1)	-
Firmicutes (35)	*Bacilli* (15)	*Bacillales* (6)	*Staphylococcaceae* (6)	*Staphylococcus* (6)
		Lactobacillales (9)	*Carnobacteriaceae* (8)	*Dolosigranulum* (8)
			Streptococcaceae (1)	*Streptococcus* (1)
	Clostridia (1)	*Clostridiales* (1)	*Peptococcaceae* (1)	*Desulfosporosinus* (1)
	Mollicutes (18)	*Incertae_sedis_8* (18)	*Erysipelotrichaceae* (18)	-
Proteobacteria (67)	*Alphaproteobacteria* (18)	*Rhizobiales* (12)	*Methylobacteriaceae* (12)	*Methylobacterium* (12)
		Rhodospirillales (2)	*Rhodospirillaceae* (2)	*Azospirillum* (2)
		Sphingomonadales (3)	*Sphingomonadaceae* (3)	*Sphingomonas* (1)
	Betaproteobacteria (24)	Burkholderiales (2)	*Alcaligenaceae* (1)	-
			Oxalobacteraceae (1)	*Massilia* (1)
		Methylophilales (2)	*Methylophilaceae* (2)	*Methylophilus* (2)
		Nitrosomonadales (1)	*Nitrosomonadaceae* (1)	*Nitrosomonas* (1)
		Rhodocyclales (18)	*Rhodocyclaceae* (18)	*Azonexus* (12)
				Sterolibacterium (1)
	Deltaproteobacteria (1)	*Bdellovibrionales* (1)	*Bdellovibrionaceae* (1)	*Bdellovibrio* (1)

(continued)

Table 4.9 (Continued)

Phylum	Class	Order	Family	Genus
	Gammaproteobacteria (24)	*Enterobacteriales* (9)	*Enterobacteriaceae* (9)	*Shigella* (3)
		Pseudomonadales (13)	*Pseudomonadaceae* (13)	*Pseudomonas* (13)
		Xanthomonadales (1)	*Xanthomonadaceae* (1)	*Stenotrophomonas* (1)
Spirochaetes (1)	*Spirochaetes* (1)	*Spirochaetales* (1)	*Spirochaetaceae* (1)	-

*For each taxa, the number of sequences identified at a 95% confidence level is shown in parenthesis.

†A total of 184 Bacteria sequences were obtained from the Prospect Tank at Louisville; 183 were classified at the phylum level, 166 at the class level, 162 at the order level, 159 at the family level, and 123 at the genus level.

Table 4.10
Published sequences perfectly matching clones sequenced in this study

Name*	Matching clone[†]	Classification (phylum, class, and genus)[‡]	Environment	Reference title	GenBank accession number	Reference
Corynebacterium accolens	Fin_10/06, Mar_10/06	*Actinobacteria*/ *Actinobacteria*/ *Corynebacterium*	Clinical or food	Taxonomic characterization of nine strains isolated from clinical and environmental specimens, and proposal of *Corynebacterium tuberculostearicum* sp. nov	AJ439346	(Feurer et al. 2004)
Gordonia sp. DEOB200	Chl+_12/05	*Actinobacteria*/ Actinobacteria/ *Gordonia*		Physiological, numerical and molecular characterization of alkyl ether-utilizing rhodococci	AY927227	(Kim et al. 2007)
uncultured actinobacterium	Pro_09/05	*Actinobacteria*/ *Actinobacteria*/ *Microcella*	Fresh water	Actinobacterial 16S rRNA genes from freshwater habitats cluster in four distinct lineages	AJ575501	(Warnecke et al. 2004)
Mycobacterium sp. M0183	UWL_01/07	*Actinobacteria*/ Actinobacteria/ *Mycobacterium*	Unknown	-	AF055332	(unp.)
Mycobacterium fuerthensis	LO3_04/07, LO3_10/06, Chl+_03/06	*Actinobacteria*/ *Actinobacteria*/ *Mycobacterium*	Patient wound	Isolation of novel bacterial species *Mycobacterium fuerthensis*	AF316618	(unp.)
Mycobacterium ratisbonense	Chl-_03/06	*Actinobacteria*/ *Actinobacteria*/ *Mycobacterium*	Sewage treatment plant	Microbial degradation of the multiply branched alkane 2,6,10,15,19, 23-hexamethyltetracosane (Squalane) by *Mycobacterium fortuitum* and *Mycobacterium ratisbonense*	AJ271863	(Berekaa and Steinbuchel 2000)

(continued)

Table 4.10 (Continued)

Name*	Matching clone[†]	Classification (phylum, class, and genus)[‡]	Environment	Reference title	GenBank accession number	Reference
Mycobacterium sp. HXN1500	Mar_10/06, Chl+_12/05, Chl-_09/05	*Actinobacteria*/ *Actinobacteria*/ *Mycobacterium*	Unknown	Alkane hydroxylases in Gram-positive strains	AJ457057	(unp.)
Mycobacterium gordonae	Chl+_12/05	*Actinobacteria*/ *Actinobacteria*/ *Mycobacterium*	Drinking water distribution system or clinical specimen	Isolation and automated ribotyping of *Mycobacterium lentiflavum* from drinking water distribution system and clinical specimens	AJ581472	(Tsitko et al. 2006)
Actinobacterium PI_GH1.1.B5	LO1_10/06, LO1_12/06, LO3_04/07, LO3_10/06, Pro_05/06, UWH_01/07, UWH_08/05, UWH_11/05, UWL_01/07, UWL_11/06	*Actinobacteria*/ *Actinobacteria*/ *Mycobacterium*	Soil	Cultivating the uncultured	AY162040	(Zengler et al. 2002)
Mycobacterium sacrum	Mar_05/06, Mar_09/05	*Actinobacteria*/ *Actinobacteria*/ *Mycobacterium*	Holy water fountain	Description of '*Mycobacterium sacrum*' sp. nov.	AY235429	(unp.)
Mycobacterium gordonae	Chl+_12/05	*Actinobacteria*/ *Actinobacteria*/ *Mycobacterium*	Frog body	*Mycobacterium gordonae* Infection in a Colony of African Tropical Clawed Frogs (*Xenopus tropicalis*)	EF428556	(unp.)

(continued)

Table 4.10 (Continued)

Name*	Matching clone[†]	Classification (phylum, class, and genus)[‡]	Environment	Reference title	GenBank accession number	Reference
Propionibacterium acnes	Fin_10/06	*Actinobacteria/ Actinobacteria/ Propionibacterium*	Blood component	*Propionibacterium acnes* 16S ribosomal RNA gene, partial sequence	AB108484	(unp.)
Nocardiaceae bacterium KVD-1982-10	Chl+_04/05	*Actinobacteria/ Actinobacteria/ Rhodococcus*	Volcanic deposits	Representation of heterotrophic isolates in respective clone libraries from recent Hawaiian volcanic deposits (<300 yr old)	DQ490426	(unp.)
Flavobacterium sp. YO66	Chl-_03/06, Chl-_09/05	*Bacteroidetes/ Flavobacteria/ Flavobacterium*	Aquaculture system	Phenotypic and molecular characterisation of fish-borne *Flavobacterium johnsoniae*-like isolates from aquaculture systems in South Africa	DQ778318	(Flemming et al. 2007)
uncultured bacterium	Mar_05/06	*Bacteroidetes/ Sphingobacteria/ Arcicella*	Groundwater	Impacts on microbial communities and cultivable isolates from groundwater contaminated with high levels of nitric acid-bearing uranium waste at the NABIR-FRC	AY661977	(unp.)
uncultured bacterium	Chl+_04/05	*Bacteroidetes/ Sphingobacteria/ Niastella*	Avian gut	Examination of the microbial ecology of the avian intestine in vivo using bromodeoxyuridine	EF025210	(Scupham 2007)

(continued)

Table 4.10 (Continued)

Name*	Matching clone[†]	Classification (phylum, class, and genus)[‡]	Environment	Reference title	GenBank accession number	Reference
Flavobacteria bacterium KF030	Chl+_03/06	*Bacteroidetes/ Sphingobacteria/ Terrimonas*	Fresh water	Bacterioplankton in shallow eutorophic lake in Japan	AB269814	(unp.)
Brevibacillus thermoruber	Pro_03/07	*Firmicutes/ Bacilli/ Brevibacillus*	Unknown	Cloning and characterization of Lon protease gene from thermophilic isolate *Brevibacillus thermoruber* WR-249	AY196006	(unp.)
uncultured bacterium	Mar_10/06	*Firmicutes/ Bacilli/ Dolosigranulum*	Spacecraft factory	Molecular bacterial community analysis of clean rooms where spacecraft are assembled	DQ532156	(Moissl et al. 2007)
uncultured bacterium	Mar_05/06	*Firmicutes/ Bacilli/ Staphylococcus*	Gastrointestinal tract	Culture-independent analysis of gut bacteria: the pig gastrointestinal tract microbiota revisited	AF371538	(Leser et al. 2002)
Staphylococcus epidermidis RP62A	Fin_10/06	*Firmicutes/ Bacilli/ Staphylococcus*	Biofilm	Insights on Evolution of Virulence and Resistance from the Complete Genome Analysis of an Early Methicillin-Resistant *Staphylococcus aureus* Strain and a Biofilm-Producing Methicillin-Resistant *Staphylococcus epidermidis* Strain	CP000029	(Gill et al. 2005)

(continued)

Table 4.10 (Continued)

Name*	Matching clone[†]	Classification (phylum, class, and genus)[‡]	Environment	Reference title	GenBank accession number	Reference
uncultured *Bradyrhizobium* sp.	Chl-_03/06	*Proteobacteria/ Alphaproteobacteria/ Afipia*	Spacecraft	Microbial characterization of the Mars Odyssey spacecraft and its encapsulation facility	AF526939	(La Duc et al. 2003)
uncultured alpha proteobacterium	Chl-_09/05	*Proteobacteria/ Alphaproteobacteria/ Bradyrhizobium*	Soil	Major groups of bacteria in a pasture soil	AY395345	(unp.)
Brevundimonas sp. V4.BP.05	Chl+_09/05	*Proteobacteria/ Alphaproteobacteria/ Brevundimonas*	Marine	Das Bakterioplankton im Westlichen Mittelmeer. Analyse der taxonomischen Struktur freilebender und partikelgebundener bakterieller Lebensgemeinschaften mit mikrobiologischen und molekularbiologischen Methoden	AJ244710	(Fritz et al. 2005)
Caulobacter sp.	Chl+_09/05	*Proteobacteria/ Alphaproteobacteria/ Caulobacter*	N/A	Phylogeny and polyphasic taxonomy of *Caulobacter* species. Proposal of *Maricaulis* gen. nov. with *Maricaulis maris* (Poindexter) comb. nov. as the type species, and emended description of the genera *Brevundimonas* and *Caulobacter*	AJ227775	(Abraham et al. 1999)

(continued)

Table 4.10 (Continued)

Name*	Matching clone[†]	Classification (phylum, class, and genus)[‡]	Environment	Reference title	GenBank accession number	Reference
Caulobacter sp. BBCT11	Chl+_09/05, Chl+_12/05	*Proteobacteria/ Alphaproteobacteria/ Caulobacter*	Soil	Distribution of tetracycline and tylosin resistance genes in bacteria isolated from swine effluent impacted environments	DQ337547	(unp.)
uncultured bacterium	Chl+_09/05	*Proteobacteria/ Alphaproteobacteria/ Caulobacter*	Acid mine drainage	Diversity of 16S Ribosomal DNA-Defined Bacterial Population in Acid Mine Drainage from Shen-bu Copper Mine, China	DQ658769	(unp.)
uncultured bacterium	UWH_08/06	*Proteobacteria/ Alphaproteobacteria/ Hyphomicrobium*	Drinking water	Phylogenetic diversity of drinking water bacteria in a distribution system simulator	AY328567	(Williams et al. 2004)
uncultured bacterium	Pro_05/06	*Proteobacteria/ Alphaproteobacteria/ Methylobacterium*	Groundwater	Long-term sustainability of a high-energy, low-diversity crustal biome	DQ088797	(Lin et al. 2006)
uncultured bacterium	LO3_10/06, Mar_09/05, Pro_09/05, Pro_10/06, UWH_01/07	*Proteobacteria/ Alphaproteobacteria/ Methylocapsa*	Groundwater	Long-term sustainability of a high-energy, low-diversity crustal biome	DQ088783	(Lin et al. 2006)
Novosphingobium subterraneum	Chl+_04/05	*Proteobacteria/ Alphaproteobacteria/ Novosphingobium*	Marine	Three species isolated from deep-sea sediments	AB025014	(unp.)

(continued)

Table 4.10 (Continued)

Name*	Matching clone[†]	Classification (phylum, class, and genus)[‡]	Environment	Reference title	GenBank accession number	Reference
Caulobacter sp. A1	UWL_11/06	*Proteobacteria/ Alphaproteobacteria/ Phenylobacterium*	Fresh water	Predator-specific enrichment of *Actinobacteria* from a cosmopolitan freshwater clade in mixed continuous culture	AF361188	(Pernthaler et al. 2001)
Antarctic bacterium R-9216	UWL_08/05	*Proteobacteria/ Alphaproteobacteria/ Porphyrobacter*	Microbial mat	Diversity of 746 heterotrophic bacteria isolated from microbial mats from ten Antarctic lakes	AJ441007	(Van Trappen et al. 2002)
uncultured bacterium	UWH_08/06, UWH_11/05	*Proteobacteria/ Alphaproteobacteria/ Porphyrobacter*	Drinking water	Population diversity in model drinking water biofilms receiving chlorine or monochloramine residual	AY957902	(Williamas et al., unp.)
uncultured bacterium	UWH_10/06	*Proteobacteria/ Alphaproteobacteria/ Sandarakinorhabdus*	Drinking water	Population diversity in model drinking water biofilms receiving chlorine or monochloramine residual	AY957890	(Williams et al., unp.)
uncultured bacterium	LO1_10/06, UWH_10/06, UWL_01/07	*Proteobacteria/ Alphaproteobacteria/ Sphingomonas*	Drinking water	Phylogenetic diversity of drinking water bacteria in a distribution system simulator	AY328556	(Williams et al. 2004)
Sphingopyxis sp. SM105	Chl+_09/05	*Proteobacteria/ Alphaproteobacteria/ Sphingopyxis*	Unknown	Isolation and characterization of *Sphingopyxis* sp. nov. strain AG13	EF424405	(unp.)

(continued)

Table 4.10 (Continued)

Name*	Matching clone[†]	Classification (phylum, class, and genus)[‡]	Environment	Reference title	GenBank accession number	Reference
Acidovorax sp. BSB421	Chl-_03/06, Chl-_09/05	*Proteobacteria/ Betaproteobacteria/ Acidovorax*	Activated sludge	Genotypic diversity of *Acidovorax* strains isolated from activated sludge and description of *Acidovorax defluvii* sp. nov	Y18617	(Schulze et al. 1999)
uncultured bacterium	Pro_09/05	*Proteobacteria/ Betaproteobacteria/ Denitratisoma*	Drinking water	Phylogenetic diversity of drinking water bacteria in a distribution system simulator	AY328709	(Williams et al. 2004)
uncultured rape *rhizosphere bacterium* wr0017	Chl-_03/06	*Proteobacteria/ Betaproteobacteria/ Limnobacter*	Plant rhizoplane	Phylogenetic Analysis of Microbial Diversity in the Rhizoplane of Oilseed Rape (Brassica napus cv. Westar) Employing Cultivation-Dependent and Cultivation-Independent Approaches	AJ295478	(Kaiser et al. 2001)
Limnobacter sp. HTCC376	UWH_08/05, UWL_09/06	*Proteobacteria/ Betaproteobacteria/ Limnobacter*	Groundwater	Bacterial community composition determined by culture-independent and -dependent methods during propane-stimulated bioremediation in trichloroethene-contaminated groundwater	AY429718	(Connon et al. 2005)

(continued)

Table 4.10 (Continued)

Name*	Matching clone[†]	Classification (phylum, class, and genus)[‡]	Environment	Reference title	GenBank accession number	Reference
uncultured bacterium	UWL_09/06	*Proteobacteria/ Betaproteobacteria/ Limnobacter*	Groundwater	Comparison of Perchloroethylene (PCE) Dechlorination Activities by Performing Microcosm Tests Using Groundwater from Four Different PCE-contaminated Aquifers in Korea	AY625137	(unp.)
uncultured bacterium	Chl-_03/06	*Proteobacteria/ Betaproteobacteria/ Limnobacter*	Marine	Study of the diversity of marine microorganisms of Coco's Island based on environmental DNA samples	EF574524	(unp.)
Acinetobacter sp. phenon 7	Chl-_09/05	*Proteobacteria/ Gammaproteobacteria/ Acinetobacter*	Unknown	New *Acinetobacter* species	AJ303013	(unp.)
Acinetobacter sp. phenon 7	Chl-_03/06, Chl-_09/05	*Proteobacteria/ Gammaproteobacteria/ Acinetobacter*	Unknown	*Acinetobacter* sp. phenon 7	AJ626712	(unp.)
Escherichia coli	Mar_05/06	*Proteobacteria/ Gammaproteobacteria/ Escherichia*	N/A	DNA sequence and analysis of 136 kilobases of the Escherichia coli genome: organizational symmetry around the origin of replication	L10328	(Burland et al. 1993)

(continued)

Table 4.10 (Continued)

Name*	Matching clone[†]	Classification (phylum, class, and genus)[‡]	Environment	Reference title	GenBank accession number	Reference
Pseudomonas putida	Mar_03/07	*Proteobacteria*/ *Gammaproteobacteria*/ *Pseudomonas*	Unknown	The conjugative bph-sal element is widely distributed among various biphenyl-utilizing bacteria	AB109776	(unp.)
Pseudomonas sp.	Mar_03/07	*Proteobacteria*/ *Gammaproteobacteria*/ *Pseudomonas*	Insect gut	Intergeneric transfer of conjugative and mobilizable plasmids harbored by *Escherichia coli* in the gut of the soil microarthropod *Folsomia candida* (Collembola)	AJ002813	(Hoffmann et al. 1998)
Pseudomonas fluorescens	Mar_10/06	*Proteobacteria*/ *Gammaproteobacteria*/ *Pseudomonas*	Plant stem	New strains of Grapevine-Associated Bacteria Inducing Leaf Protection Against Botrytis cinerea and Defense-Related Reactions	AM293679	(unp.)
Pseudomonas putida	Chl+_09/05	*Proteobacteria*/ *Gammaproteobacteria*/ *Pseudomonas*	Soil	Plasmid-mediated degradation of 4-chloronitrobenzene by newly isolated *Pseudomonas putida* strain ZWL73	AY647158	(Zhen et al. 2006)
gamma *proteobacterium* GPTSA100-22	Fin_05/06	*Proteobacteria*/ *Gammaproteobacteria*/ *Rheinheimera*	Spring water	Study of culturable bacterial diversity of an aquatic sample collected from a warm spring, Assam, India	DQ914845	(unp.)

(continued)

Table 4.10 (Continued)

Name*	Matching clone†	Classification (phylum, class, and genus)‡	Environment	Reference title	GenBank accession number	Reference
Escherichia coli	Mar_05/06	*Proteobacteria/ Gammaproteobacteria/ Shigella*	N/A	Sequence heterogeneities among 16S ribosomal RNA sequences, and their effect on phylogenetic analyses at the species level	X80728	(Cilia et al. 1996)
Stenotrophomonas sp. IC193	Chl-_09/05	*Proteobacteria/ Gammaproteobacteria/ Stenotrophomonas*		Diversity of carbazole-degrading bacteria having the car gene cluster: isolation of a novel gram-positive carbazole-degrading bacterium	AB196256	(Inoue et al. 2005)
bacterium 3C10	Chl-_09/05	*Proteobacteria/ Gammaproteobacteria/ Stenotrophomonas*	Lab reactor	Effect of nutrient periodicity on microbial community dynamics	DQ298778	(Carrero-Colon et al. 2006)

*Name given in GenBank entry.
†Coded based on chloraminated environment where clone was retrieved from and date.
‡According to RDP Release 9 (Cole et al. 2005).

DETERMINATION OF OPERATIONAL TAXONOMIC UNITS

Partial sequences (600-900 bp in general) from the relatively less conserved 5' end of 16S rRNA were used for defining taxonomic units based on the similarity among the sequences. A cutoff of 99% similarity was applied for the definition of OTUs. Table 4.11 summarizes this analysis. From a total of 1369 good quality sequences representing the 9 different environments tested, a total of 381 OTUs were defined. Two hundred of these OTUs were represented by only one sequence in the library, and 181 of them included two or more sequences.

The diversity between each of the environments sampled was characterized using the Shannon and Simpson richness indices. As diversity increases, the Shannon index increases, while the Simpson index decreases. As shown in Table 4.11, the environments with the lowest diversity corresponded to the pilot systems operated at the Louisville Water Company, followed by the bench-scale reactor that did not receive chloramines. Other environments tested had similar diversity indices.

Table 4.11
OTU Analysis

System	Number of sequences*	Total number of OTUs†	OTUs with 2 or more sequences	Shannon Index‡	Simpson Index‡
Bench-scale reactors					
Chl+	175	52	31	3.38	0.06
Chl-	86	29	12	2.46	0.17
Combined	261	73	45	3.69	0.04
Pilot-scale DS					
UWL	221	59	32	3.41	0.05
UWH	216	59	31	3.43	0.06
LO1	92	12	6	1.62	0.31
LO3	86	11	7	1.30	0.43
Combined	615	115	61	3.60	0.08
Full-scale DS					
Finished water	132	54	22	3.60	0.03
Martin's Hill Tank	185	58	19	3.35	0.06
Prospect Tank	176	47	21	3.20	0.06
Combined	493	135	56	4.19	0.02

*number of sequences after removing chimeras and misaligned sequences.
†Analysis done for a similarity cutoff of 99%.
‡Richness indexes commonly used in microbial ecology to describe the diversity of a population. As diversity increases, the Shannon index increases and the Simpson index decreases.

The combined libraries of systems at each scale were also analyzed for diversity (Table 4.11), to understand the level of overlapping between the systems. This analysis showed that diversity of unified libraries was significantly greater than individual libraries, pointing to differences between environments on the same scale. However, the number of OTUs in a combined library was always smaller than the sum of OTUs in individual libraries, demonstrating the presence of overlaps. The OTUs of combined sets were used only in this side analysis and not included in other levels of data processing.

With the definition of OTUs in individual libraries, clones from selected OTUs were processed for nearly full sequencing of the 16S rRNA in order to construct phylogenetic trees, as described in the following section.

CONSTRUCTION OF PHYLOGENETIC TREES

For a detailed description of the phylogeny of bacteria found in the reactors, phylogenic trees were constructed using the following groups that include sequences from the clone library and additional relevant sequences obtained from the public database:

1. One representative clone from each OTU with two or more sequences.
2. Public sequences that matched at least one sequence in the library with greater than 97% similarity and were reportedly retrieved from drinking water environments.
3. Sequences from OTUs with only one representative sequence in the library, only if the sequence was associated with one of the public sequences selected in 2.
4. Type strains most similar to OTU representatives selected in 1 and 3, provided that the similarity was greater than 90%.
5. One or two sequences from the public database that matched OTU representative sequences from 1 and 3 if no closely associated (>95%) type strain sequence was found in 4. The similarity cutoff for these sequences was set to greater than 95% similarity.
6. Type strains representing known pathogens that are associated with the selected OTUs at the genus level.

Nearly full sequences (i.e., sequence length greater than 1,200 bp) were obtained from most selected clones for tree construction. Sixteen of the selected clones could not be sequenced from the 3' end, resulting in a total of 194 nearly full library sequences placed in the phylogenetic trees. Also, public sequences were accepted only if they were greater than 1,200 bp long.

The resulting phylogenetic trees are presented in Figures 4.2 to 4.7. In these figures, publicly available sequences are indicated by their GenBank accession codes, references are provided for published sequences obtained from drinking water environments, and the sequences from the library created in this project are named based on the system they were retrieved and the OTU that they belong to. The numbers in parentheses indicate number of sequences in the corresponding OTU and T denotes that the sequence correspond to a type strain.

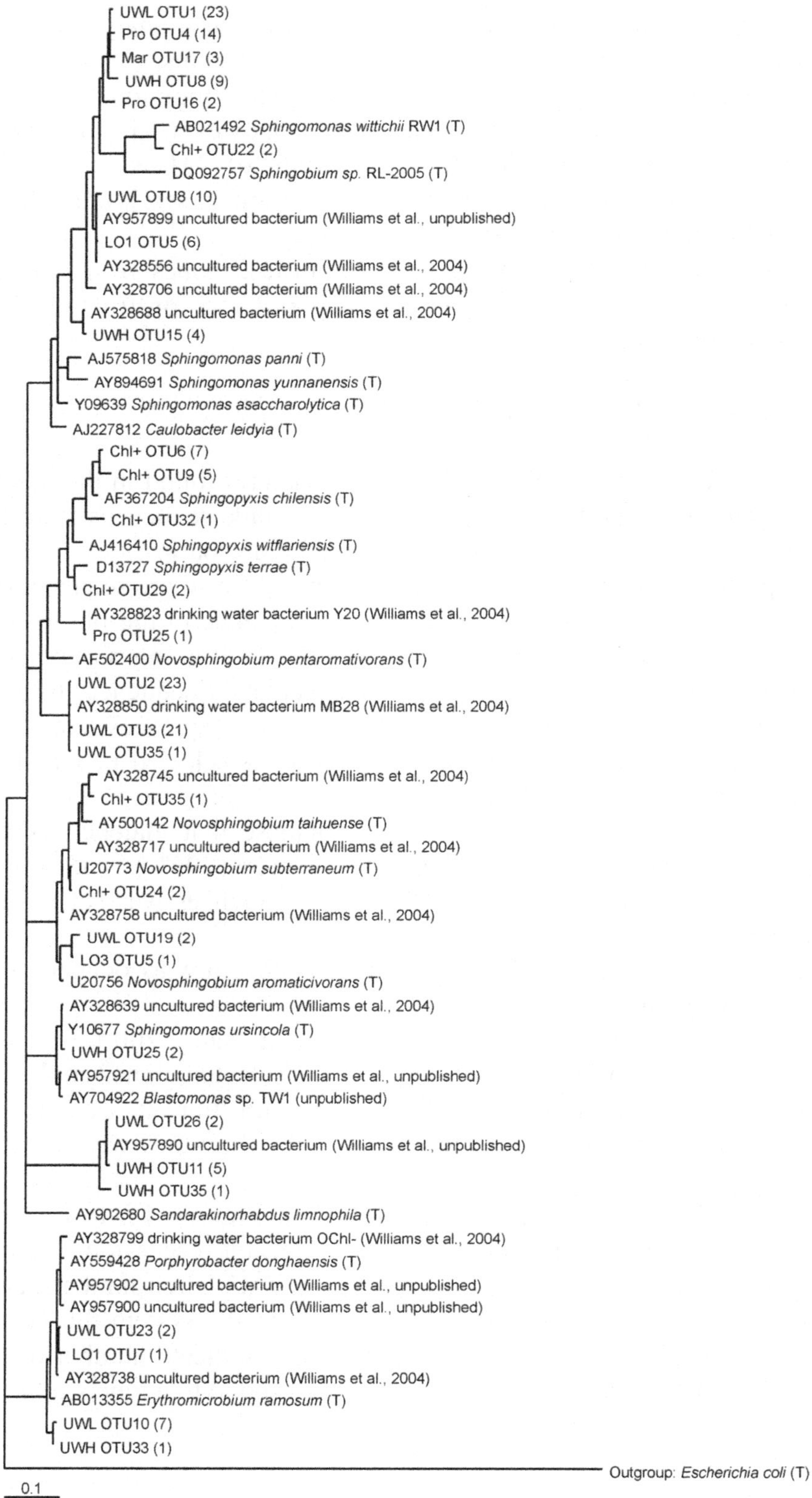

Figure 4.2 Phylogenetic trees of the *Sphingomonadales* family within the *Alphaproteobacteria*.

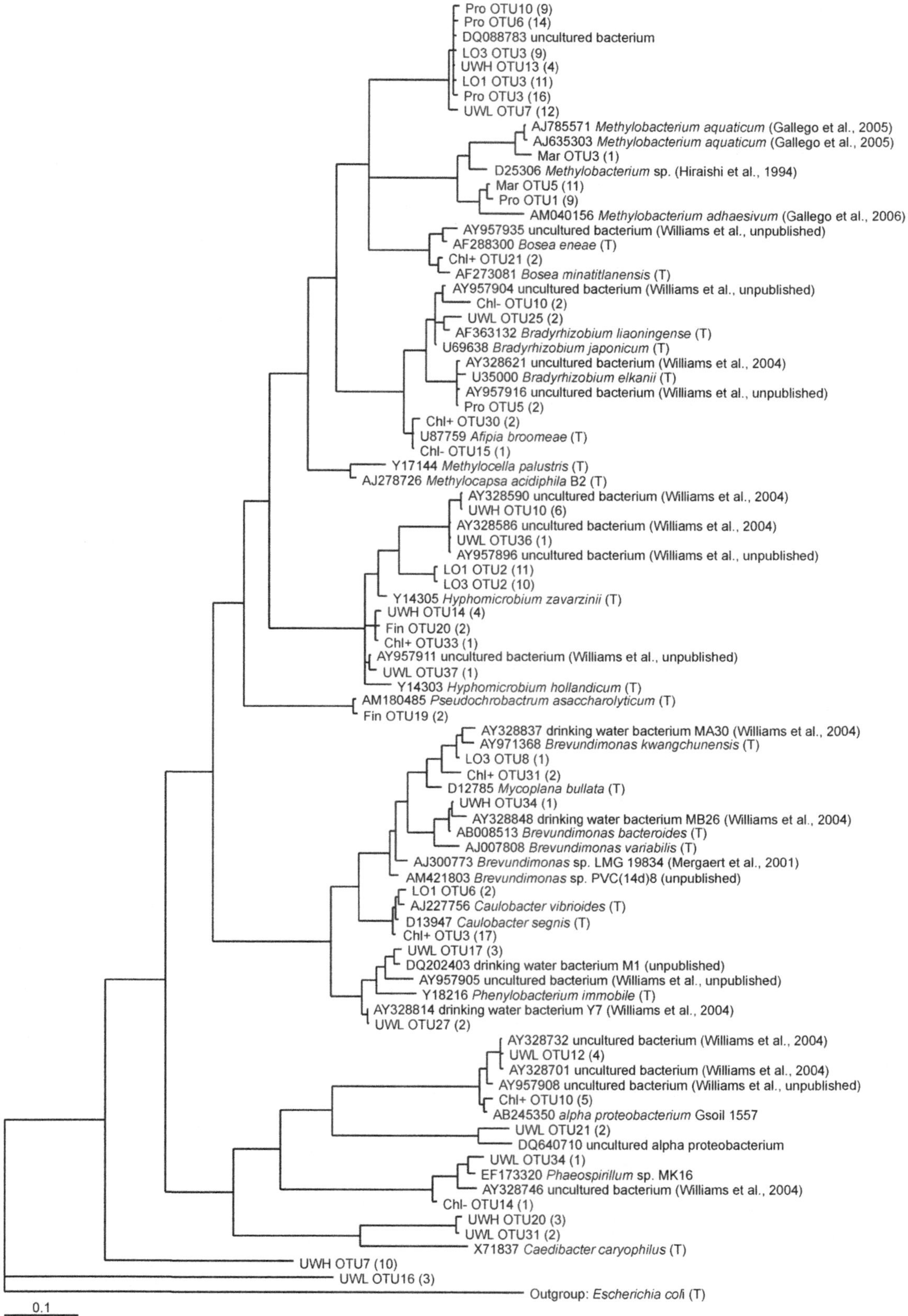

Figure 4.3 Phylogenetic tree of the *Alphaproteobacteria*, excluding the *Sphingomonadales* family.

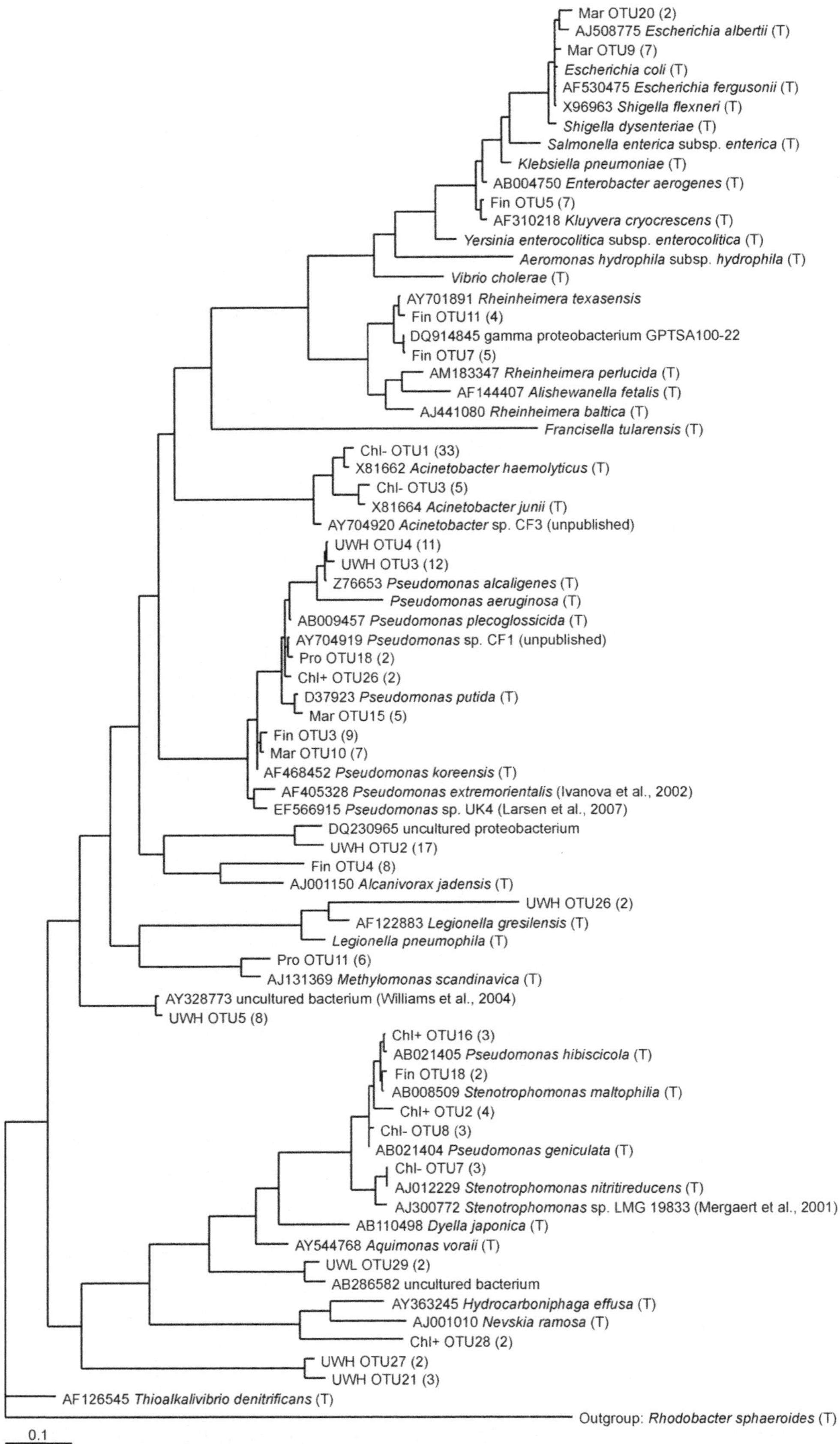

Figure 4.4 Phylogenetic tree of sequences belonging to the *Gammaproteobacteria*.

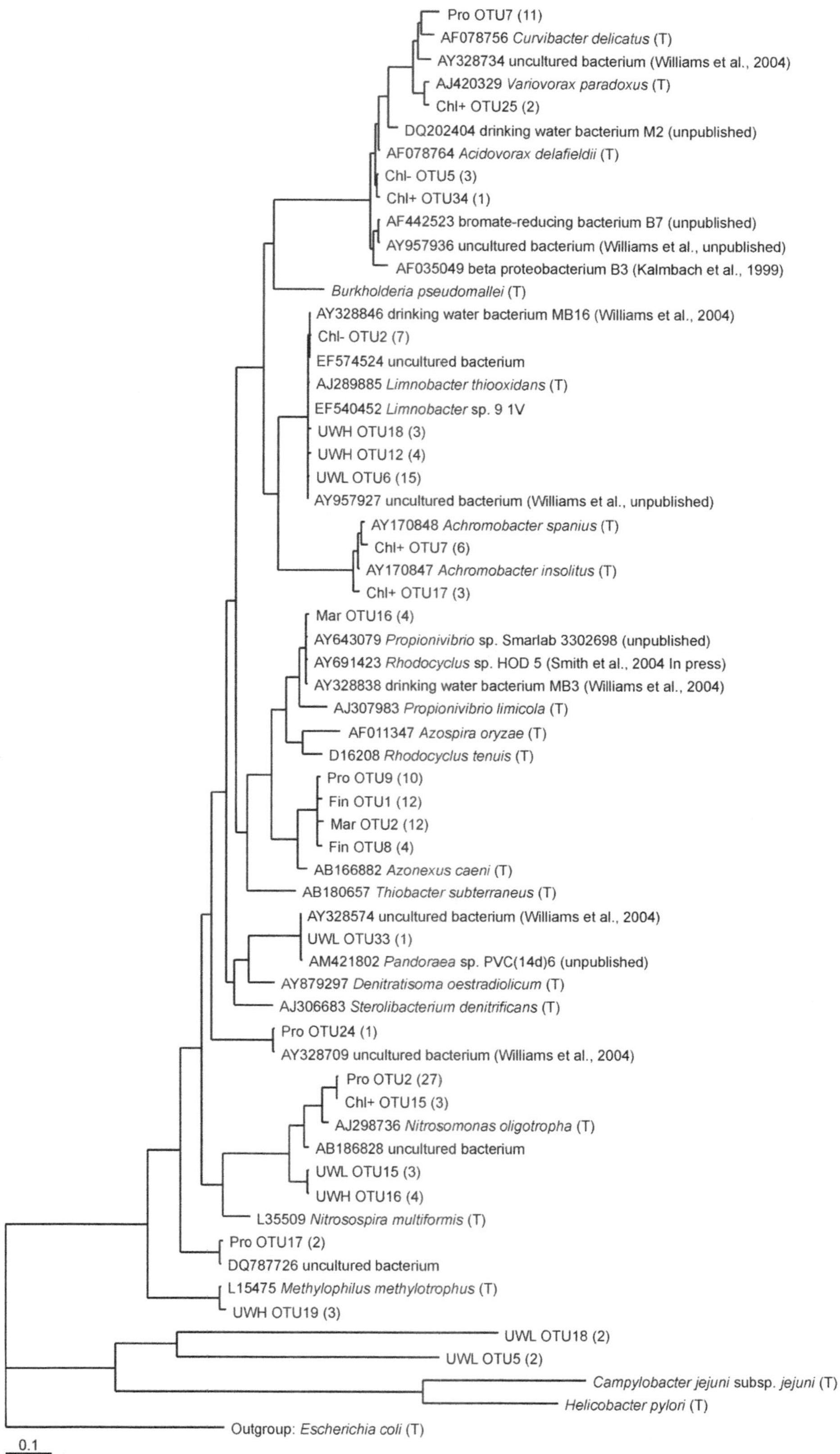

Figure 4.5 Phylogenetic tree of sequences related to *Proteobacteria*, excluding *Alpha-* and *Gammaproteobacteria*.

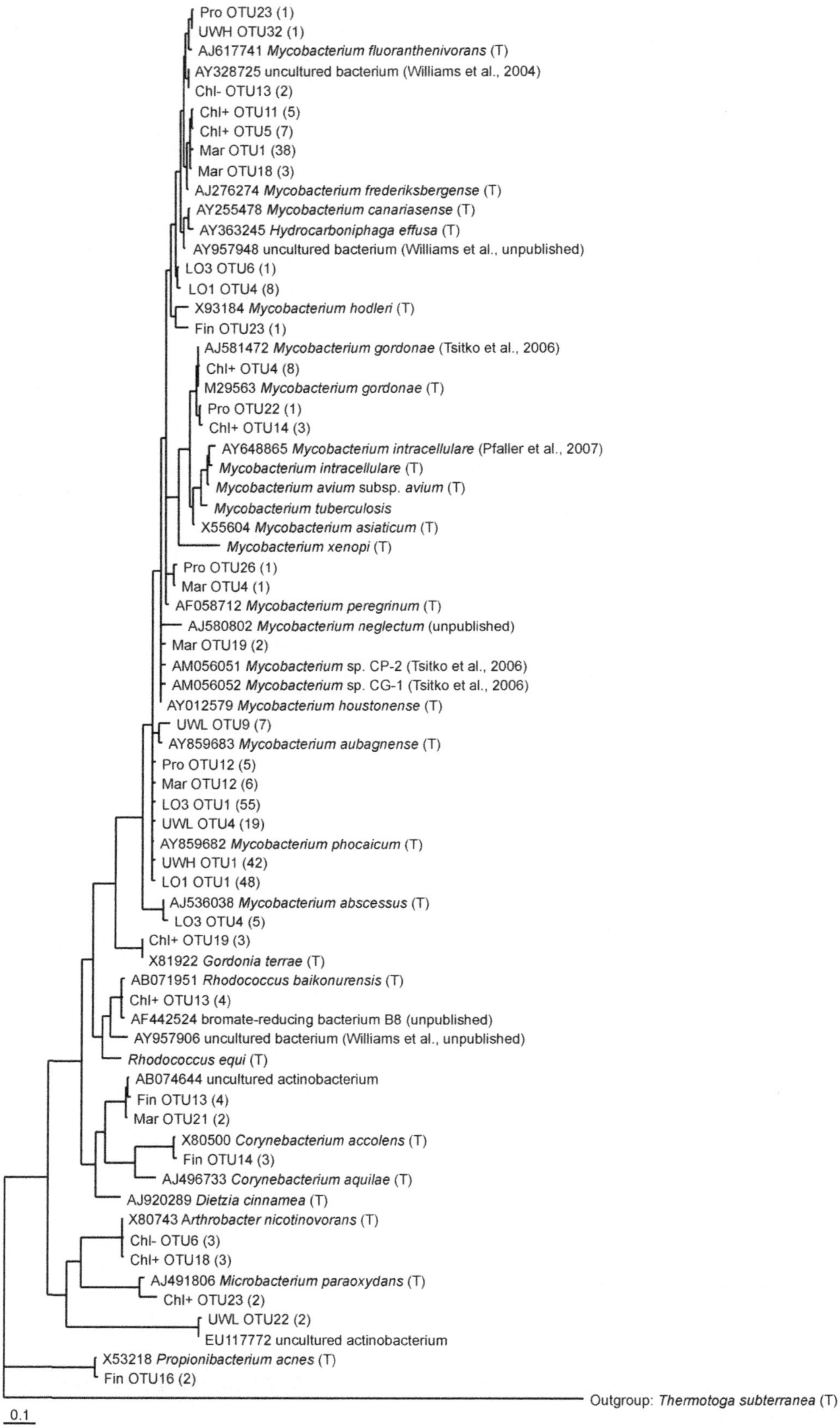

Figure 4.6 Phylogenetic tree of sequences related to the phylum *Actinobacteria*.

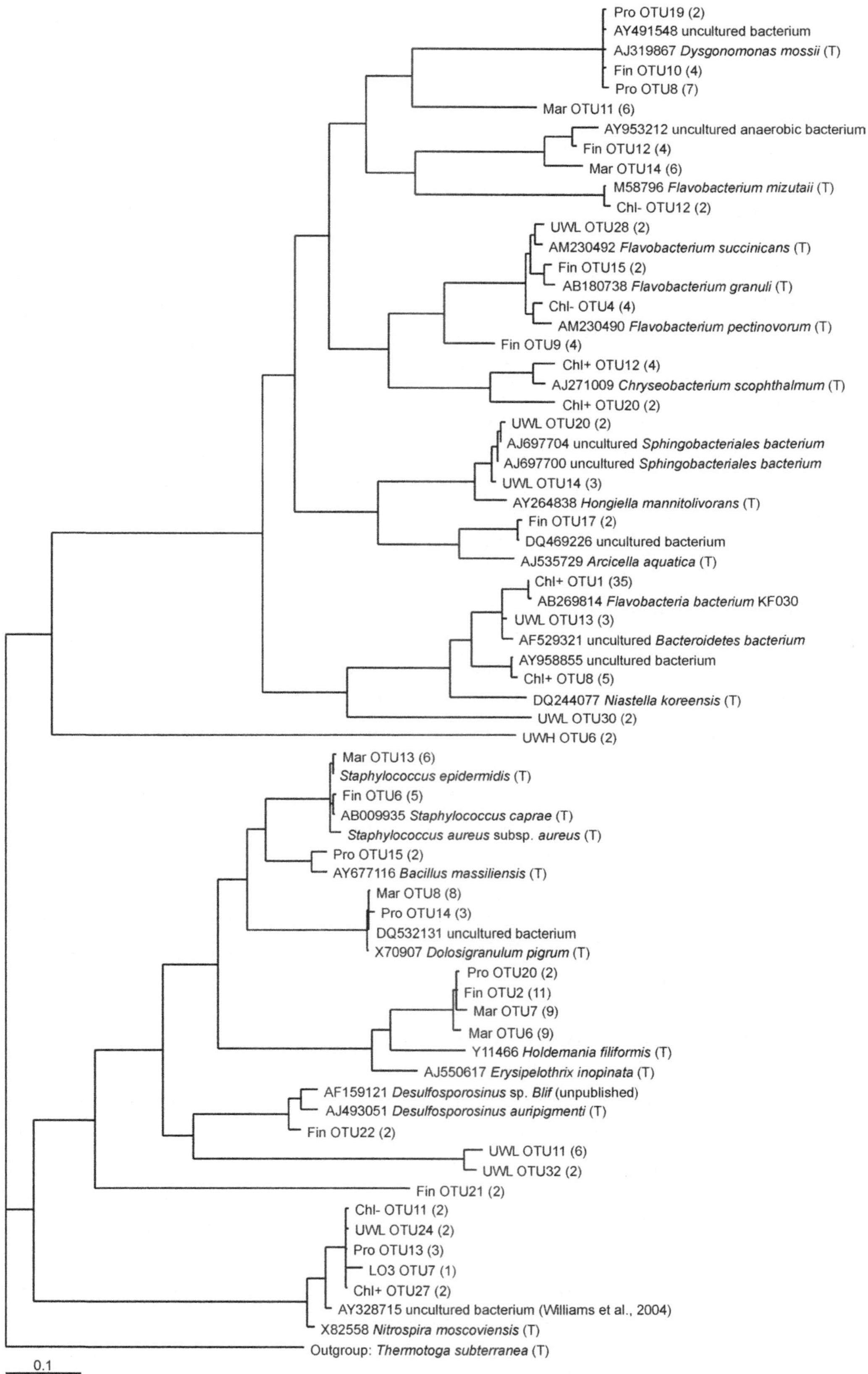

Figure 4.7 Phylogenetic tree of Bacteria sequences other than *Proteobacteria* and *Actinobacteria*.

COMPARISONS AMONG THE DIFFERENT SYSTEMS STUDIED

One of the objectives of this study was to determine if bench- and pilot-scale chloraminated systems are adequate models of the real scale systems. The phylogenetic analyses provided substantial information about the similarity of communities in the different scales studied.

As discussed above, taxonomic distribution of partial sequences were similar at phylum level, but significant differences, as well as some similarities, were observed at lower taxa (Tables 4.1-4.9). A striking observation was the presence of the *Mycobacterium* genus and *Sphingomonadales* family at all scales. Within-scale differences provided evidence for the association of these two groups with chloramination. Community compositions in bench-scale reactors suggested both *Mycobacterium* and *Sphingomonadales* were selected for upon chloramination (Tables 4.1 and 4.2). Consistently, these organisms were ubiquitous in full-scale distribution system tanks (Tables 4.8 and 4.9), but rare or absent in finished water at the beginning of the distribution system. In contrast, *Mycobacterium* was dominant in both the non-chloraminated (LO1) and chloraminated (LO3) simulators at Louisville, and *Sphingomonadales* seemed more abundant in the non-chloraminated system (Tables 4.5 and 4.6). However, the non-chloraminated simulator was sampled in the first three months of the run, which may not be a long enough period to see the possible elimination of these subpopulations from the original seed. Note that the seed was obtained from the Chl+ reactor (see Chapter 2), and therefore, may have contained *Mycobacterium*- and *Sphingomonadales*-related organisms (see Table 4.2), although a molecular analysis of the seeding community was not carried out. Furthermore, LO3 was fed with chlorite, making it difficult to evaluate the effect of chloramination alone. Thus, we hypothesize that chloraminated environments support the growth of *Mycobacterium*- and *Sphingomonadales*-related organisms, although firm conclusions cannot be reached at present.

Phylogenetic trees also showed interesting relationships between sequences retrieved from the different scales. Each phylogenetic group represented in Figures 4.2 to 4.7 included sequences from all three scales, once again suggesting that the microbial communities of the simulators and full-scale systems were similar at higher levels of taxa. However, lower orders of clustering in these trees showed opposite trends in terms of the relatedness of organisms from different scales. There were clusters of mixed scales suggesting bench-scale chemostats and pilot-scale simulators were able to grow organisms inhabiting full-scale distribution systems. Two such clusters were the *Nitrospira* lineage in Figure 4.7, and a *Nitrosomonas*-related branch in Figure 4.5, which represent NOB and AOB, respectively, as previously reported (Regan et al. 2002; Regan et al. 2003), and indicate the presence of common nitrifiers in the different systems. More important for the current study are the mixed-scale clusters of heterotrophs, including but not limited to multiple subclusters under the *Sphingomonadales* family (Figure 4.2), a *Pseudomonas* cluster (Figure 4.4), a *Limnobacter* cluster (Figure 4.5), and multiple *Mycobacterium* subclusters (Figure 4.6). Some of these extend to other studied distribution systems, and in a way, form established drinking water clusters (e.g. *Limnobacter* cluster in Figure 4.5). However, there are also lineages involving only full-scale distribution system clones such as the *Azonexus* (Figure 4.5) and *Staphylococcus* (Figure 4.7) clusters. The opposite is also observed. For example the *Arthrobacter* cluster (Figure 4.6) involves only bench-scale chemostat clones, suggesting that *Arthrobacter*-related species may not be relevant to full-scale systems. These observations imply that bench- and pilot-scale systems may introduce significant biases in the simulation of heterotrophic communities in distribution systems. However, another possible explanation for the presence of organisms not common to all scales is the influence of system-specific factors such as the source water.

In conclusion, whether bench- and pilot-scale chloraminated systems can adequately represent the actual scale is questionable since there are significant differences between microbial populations in different systems. In this study, systems at the three different scales had different source waters, which might have affected the microbial community composition due to differences in water quality or the microbial seed. Therefore, the effect of source water must be systematically studied to fully evaluate the representative ability of distribution system simulators. However it is clear that at least close relatives of some organisms in full-scale systems can be grown in bench- or pilot-scale systems, indicating that simulation studies are valuable for understanding the effects of disinfection strategies on microbial subpopulations. Chapter 5 presents further evaluations of differences between populations in the different systems studied, from the point of view of statistical analyses of community fingerprints.

CHAPTER 5
RESULTS AND DISCUSSION: ARISA COMMUNITY FINGERPRINTS

In Chapter 4, the community composition of the chloraminated environments studied was described based on the diversity of 16S rRNA sequences retrieved from the samples. In this chapter, community profiles are characterized with respect to the ITS fragment length, using ARISA electropherograms. The details of the method were described in Chapter 3. In addition to offering independent molecular information on community composition, the ARISA fingerprinting technique has two important advantages. First, while the 16S sequencing method relies on sampling of a limited set of clones, ARISA electropherograms provide information from all detectable organisms in a sample, within a single profile. Second, since ARISA is time-efficient and requires much less resources than sequencing, it allows the analysis of a larger number of samples with a smaller budget. Using these advantages, clustering analyses were carried out for a sufficient number of ARISA fingerprints, each representing the full community profile at a given time point in a given system, to achieve a statistically reliable comparison of microbial communities for the different systems studies. Moreover, if the ITS fragment lengths can be matched to 16S sequences in a clone library (i.e. using the clone-ARISA technique), ARISA can potentially provide full phylogenetic information on samples, once a comprehensive library of 16S-ITS clones is obtained. This option was also fully investigated with the current 16S-ITS database from the distribution systems studied.

The distribution system samples profiled with ARISA are shown in Figure 5.1. The total number of samples analyzed was 8 for bench-scale chemostats, 9 for UW-based distribution system simulators, and 7 for full-scale distribution systems. Only two and three samples were available for the pilot-scale systems LO1 and LO3 established in Louisville, respectively. In full-scale finished water from Louisville, four samples did not yield detectable fragments after PCR amplification, and therefore, were excluded from the analysis.

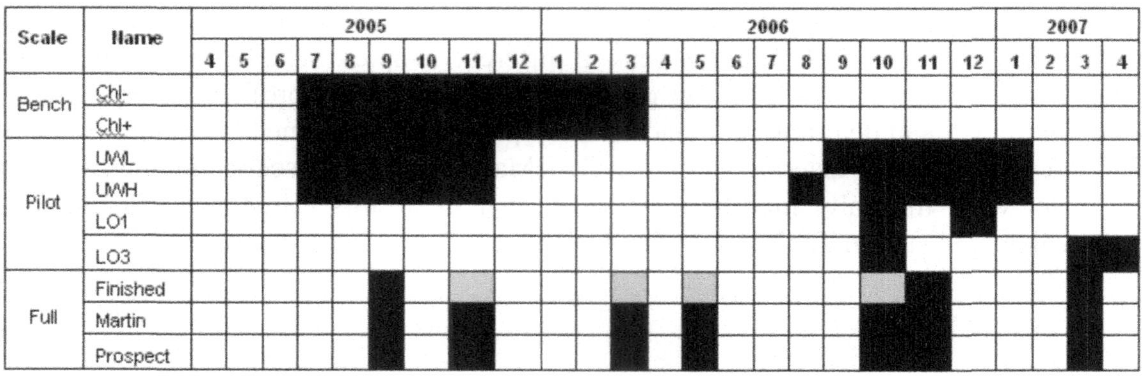

Figure 5.1 Sampling schedule for community fingerprinting. Black cells indicate successful profiling of samples at the given time points. Grey cells indicate sampling without success in fingerprinting.

In what follows, the general aspects of ARISA community profiles and clone-ARISA results are described with examples, and then, a hierarchical clustering analysis of community profiles for bench-, pilot-, and full-scale systems are separately discussed based on heat-graphs. Finally, the overall comparison of all systems studied is presented using multidimensional scaling (MDS) plots.

EXAMPLE ARISA OUTPUTS

The output of ARISA experiments is graphically presented in this section with sample electropherograms. The outputs are obtained by aligning raw fluorescence/time data with that of internal standards, which yields the lengths of all detectable fragments in a sample. There were two types of samples analyzed with ARISA, namely the community samples from the different environments studied (Figure 5.1) and clones containing the 16S-ITS insert. The former analysis is termed community ARISA and the latter clone-ARISA.

Community ARISA Profiles

ARISA profiles of selected samples are illustrated in Figure 5.2. As shown with duplicates of January 2006 samples from the bench-scale Chl+ reactor (panels a and b in Figure 5.2), replicates of the same sample resulted in very similar patterns, although the fluorescence intensity scales could be different. Figure 5.2c depicts the profile of a sample from a different, but same scale (bench) system, Chl-. Some common peaks (alleles) between the two environments are observed (those representing 848 bp and 869 bp fragment lengths). Others are present only in Chl- sample (e.g., 773 bp) or in Chl+ (e.g., 756 bp). Figure 5.2d shows the profile of a distinct sample obtained from a full-scale testing location (Prospect Tank), which does not share any of the common peaks in the two bench-scale samples, exhibits peaks that are not present in the bench-scale samples (i.e., 465 bp and 949 bp), and has a common peak with the with the Chl- sample (i.e., 784 bp). Thus, the similarity between the full-scale sample and either of the bench-scale samples would be relatively low.

It is important to note that the fragment lengths assigned to fluorescence peaks actually represent a range of measured fragment sizes (e.g. 465 represents 464.5 bp through 465.5 bp), which defines an allele associated with the given fragment size (465 bp). This is due to the slight differences in the positions of the same peak in different samples. The binning of the fragment sizes for allele definition was done by overlaying all community profiles, as exemplified by those in Figure 5.2, and determining ranges of peak positions in confined locations. Some of the alleles were automatically determined by the GeneMarker software used in fragmental analyses, while others were manually modified for most accurate results. In total, 153 alleles were defined, which are listed in Table A.1 in the Appendix section of this report.

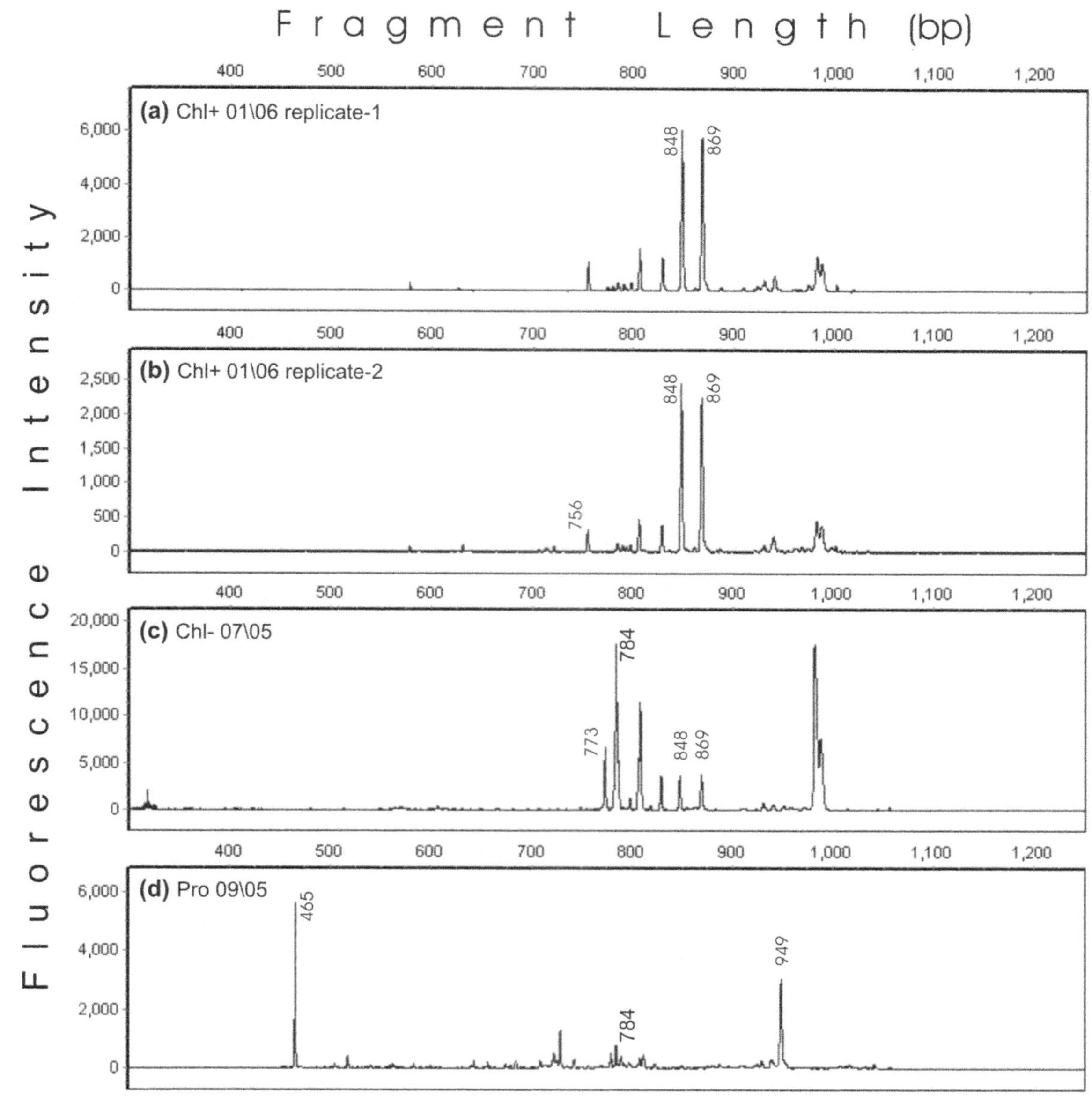

Figure 5.2 Example ARISA profiles. Peaks discussed in text are indicated.

Clone ARISA Profiles

Using the technique of clone-ARISA for matching ITS lengths to 16S signatures has been mentioned in literature (Brown et al. 2005; Newton et al. 2006), but the data is limited for the generalization of the approach with regard to successful tracking of ARISA peaks in community profiles. In this study, out of a total of nearly 1,500 16S-ITS clones from different systems, 1,250 gave well-aligned ITS peak profiles for reliable analysis. The results revealed some complications with the matching of ITS length to 16S identity, some of which were methodological, while others were due to the characteristics of the ITS alleles.

Since a 16S-ITS clone consists of a plasmid with a single fragment insert, each clone is expected to show a single peak in its ARISA profile. About 25% of the analyzed clones conformed to this theoretical expectation as exemplified in Figure 5.3a with a clone that was identified as *Arthrobacter* from its partial 16S sequence, and is associated with a single peak at 848 bp. Thus, one could assign the 16S-based phylogenetic identity of *Arthrobacter* to alleles of fragment length 848 bp observed in some community profiles (e.g., Chl+ and Chl- samples in Figure 5.2), with high confidence. However, the large majority (ca. 60%) of clone-ARISA profiles were associated with a double peak pattern, as illustrated in Figure 5.3b. The second (larger fragment) peak, which appeared generally between 80 to 105 bp to the right of the first, was eventually associated with a parallel PCR amplification in these samples because of the contamination of the DNA template with primers M13F and M13R, which were used in the amplification of DNA fragments from plasmids (see Figure 3.1). Thus, in addition to ARISA primers 1406F and 23Sr, the PCR buffer appeared to contain M13F and M13R, each of which could theoretically act as a reverse primer working with the labeled forward primer 1406F to produce detectable fragments. Since these primers target flanking plasmid DNA, they would generate a larger fragment than the true fragment obtained with 23Sr. Tests with several double-peak clones provided the proof of this concept as exemplified in Figure 5.3c, by showing that only the second peak appeared if the plasmid primers were used as the only reverse primer in PCR buffers (i.e., excluding primer 23Sr during the PCR reaction). Thus, the true peak of the clone in Figure 5.3b is 790, since the larger 881bp fragment is generated by M13R in this case. The concept of M13F or M13R contamination generating the second peak in clone-ARISA electropherograms was generalized to all samples with similar patterns as in Figure 5.3b. Consequently, the first peaks in these electropherograms were assumed to correspond to the ITS length and subsequently used for matching 16S identity to ITS length.

To acknowledge the potential complications in clone-ARISA profiles, the assignments of identity to fragment length were characterized by three qualitative levels of confidence defined as high, medium, and low as exemplified in Figure 5.3. While a single peak was always associated with the high level of confidence (Figure 5.3a, the shorter fragment in most double-peak patterns was given a medium level of confidence (Figure 5.3b). However, if the longer peak had a small intensity with respect to the shorter peak, then the level of confidence was considered high. On the contrary, if the intensity of the longer peak was greater than that of the shorter peak, then the level of confidence on peak determination was considered low (Figure 5.3d). In less than 15% of profiles, multiple peaks were observed, as shown in Figure 5.3e, where it is not clear whether 728 bp or 1012 bp should represent the ITS length. In such cases the profile was considered unreliable, unless one of the peaks was dominant, when it was generally associated with a low level of confidence. Furthermore, five ITS fragments from the genome of the host organism (i.e., competent *E. coli* cells), were occasionally detected in some clone-ARISA profiles. The size ranges for these fragments were determined and matching peaks were removed from profiles before further analysis. Although a large diversity of patterns were observed in clone-ARISA profiles, those shown in Figure 5.3 represent the vast majority.

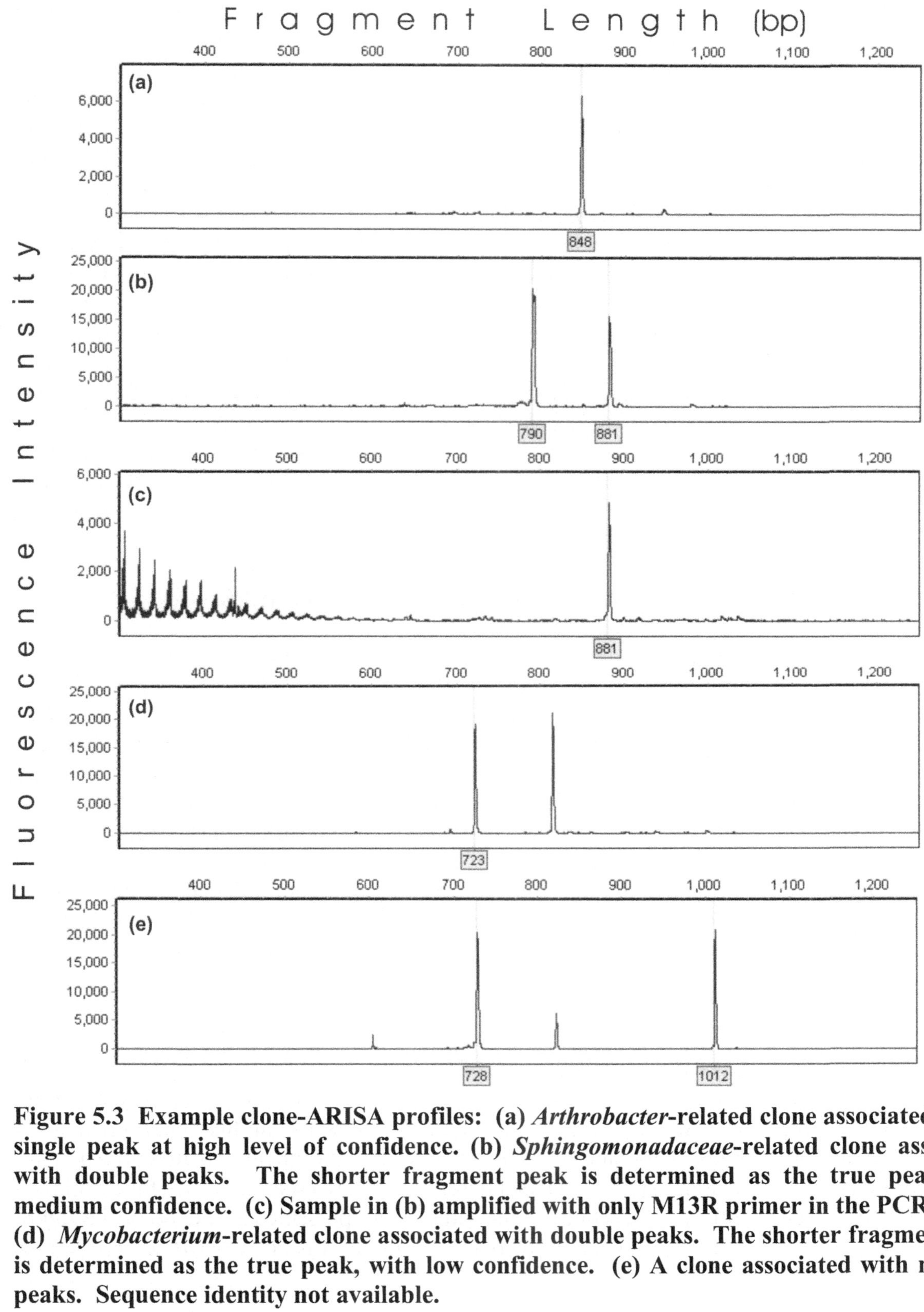

Figure 5.3 Example clone-ARISA profiles: (a) ***Arthrobacter*****-related clone associated with a single peak at high level of confidence. (b)** ***Sphingomonadaceae*****-related clone associated with double peaks. The shorter fragment peak is determined as the true peak, with medium confidence. (c) Sample in (b) amplified with only M13R primer in the PCR buffer. (d)** ***Mycobacterium*****-related clone associated with double peaks. The shorter fragment peak is determined as the true peak, with low confidence. (e) A clone associated with multiple peaks. Sequence identity not available.**

Using reliable clone-ARISA profiles, most of the 153 alleles defined based on community profiles could be matched to 16S identities, albeit at different levels of confidence. The overall matching of alleles to partial 16S sequences is shown in the Table A.1 in the Appendix. The results in Table A.1 pointed to two further complications associated with the use of the ITS gene. First, a single allele was often matched to multiple species from different taxa, including different phyla, undermining the efforts for converting ARISA fingerprints to community composition with 16S-based identities. For example, the allele 848 in Chl+ samples (Figure 5.2a) is not only associated with *Arthrobacter* within the phylum *Actinobacteria* as demonstrated above (Figure 5.3a), but also corresponded to a clone classified under *Gammaproteobacteria*, with both assignments obtained with a high level of confidence. It was also observed that ITS lengths showed great diversity at the species level. That is, members of the same genus were represented by different alleles. For example, *Caulobacter* clones were associated with alleles 953, 960, 969, 978, 1002, and 1003. This could partially be attributed to the fact that some organisms have multiple rRNA operons, and different operons can potentially have different ITS lengths. For example, pure cultures of *E. coli* show seven peaks in ARISA profiles, consistent with the presence of seven rRNA operons in *E. coli* (data not shown). In conclusion, the relationship between alleles and 16S identities is far from the ideal one-to-one correspondence, and therefore, the ARISA results presented in the following sections should be interpreted with care.

HIERARCHICAL CLUSTERING ANALYSES OF ARISA PROFILES

Standard statistical hierarchical clustering methods were used for comparative analyses of samples from the different environments studied. Before running clustering algorithms, the peaks in the ARISA community profiles were first converted to absence/presence of alleles (value of 1 for detectable presence and 0 for absence). Since samples were generally profiled at least in duplicates, the replicate profiles were then combined by taking their union set. That is, when an allele was detected in any one of the replicates, it was taken as present (1), otherwise absent (0). Using the similarities between the absence/presence profiles, clustering of samples was carried out with the Cluster software (Eisen et al. 1998; de Hoon et al. 2004). Three separate analyses were conducted, for the bench-, pilot- and full-scale environments, respectively. The outcome of these analyses is presented as pseudo-heat graphs with clustering information in two dimensions.

Bench-Scale Systems

The clustering of ARISA community profiles from the Chl+ and Chl- bench-scale reactors is depicted in Figure 5.4. Each row in the figure corresponds to a specific peak on the ARISA electropherograms, and the dark squares indicate presence of that peak on a given sample. The text at the right hand side of each row presents the allele and its corresponding identification based on the clone ARISA analysis described before (see Table A.1 for a complete list of alleles and their identity). The names of the samples are presented at the top of the figure, with information on the reactor sampled and the date the sample was taken. Alleles and samples are both clustered according to their similarity.

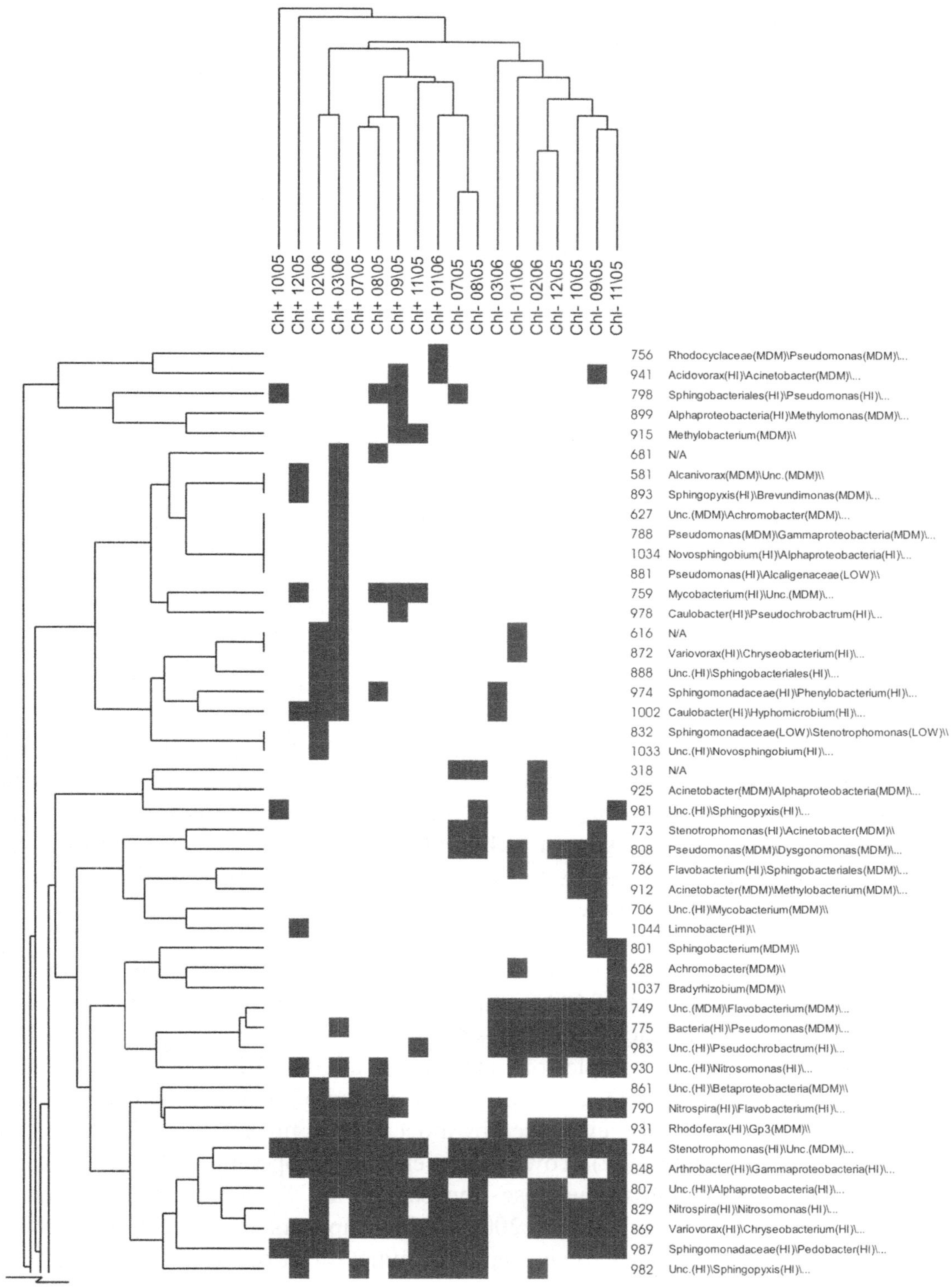
Chl+ 10\05
Chl+ 12\05
Chl+ 02\06
Chl+ 03\06
Chl+ 07\05
Chl+ 08\05
Chl+ 09\05
Chl+ 11\05
Chl+ 01\06
Chl- 07\05
Chl- 08\05
Chl- 03\06
Chl- 01\06
Chl- 02\06
Chl- 12\05
Chl- 10\05
Chl- 09\05
Chl- 11\05
756 Rhodocyclaceae(MDM)\Pseudomonas(MDM)\...
941 Acidovorax(HI)\Acinetobacter(MDM)\...
798 Sphingobacteriales(HI)\Pseudomonas(HI)\...
899 Alphaproteobacteria(HI)\Methylomonas(MDM)\...
915 Methylobacterium(MDM)\\
681 N/A
581 Alcanivorax(MDM)\Unc.(MDM)\\
893 Sphingopyxis(HI)\Brevundimonas(MDM)\...
627 Unc.(MDM)\Achromobacter(MDM)\...
788 Pseudomonas(MDM)\Gammaproteobacteria(MDM)\...
1034 Novosphingobium(HI)\Alphaproteobacteria(HI)\...
881 Pseudomonas(HI)\Alcaligenaceae(LOW)\\
759 Mycobacterium(HI)\Unc.(MDM)\...
978 Caulobacter(HI)\Pseudochrobactrum(HI)\...
616 N/A
872 Variovorax(HI)\Chryseobacterium(HI)\...
888 Unc.(HI)\Sphingobacteriales(HI)\...
974 Sphingomonadaceae(HI)\Phenylobacterium(HI)\...
1002 Caulobacter(HI)\Hyphomicrobium(HI)\...
832 Sphingomonadaceae(LOW)\Stenotrophomonas(LOW)\\
1033 Unc.(HI)\Novosphingobium(HI)\...
318 N/A
925 Acinetobacter(MDM)\Alphaproteobacteria(MDM)\...
981 Unc.(HI)\Sphingopyxis(HI)\...
773 Stenotrophomonas(HI)\Acinetobacter(MDM)\\
808 Pseudomonas(MDM)\Dysgonomonas(MDM)\...
786 Flavobacterium(HI)\Sphingobacteriales(MDM)\...
912 Acinetobacter(MDM)\Methylobacterium(MDM)\...
706 Unc.(HI)\Mycobacterium(MDM)\\
1044 Limnobacter(HI)\\
801 Sphingobacterium(MDM)\\
628 Achromobacter(MDM)\\
1037 Bradyrhizobium(MDM)\\
749 Unc.(MDM)\Flavobacterium(MDM)\...
775 Bacteria(HI)\Pseudomonas(MDM)\...
983 Unc.(HI)\Pseudochrobactrum(HI)\...
930 Unc.(HI)\Nitrosomonas(HI)\...
861 Unc.(HI)\Betaproteobacteria(MDM)\\
790 Nitrospira(HI)\Flavobacterium(HI)\...
931 Rhodoferax(HI)\Gp3(MDM)\\
784 Stenotrophomonas(HI)\Unc.(MDM)\...
848 Arthrobacter(HI)\Gammaproteobacteria(HI)\...
807 Unc.(HI)\Alphaproteobacteria(HI)\...
829 Nitrospira(HI)\Nitrosomonas(HI)\...
869 Variovorax(HI)\Chryseobacterium(HI)\...
987 Sphingomonadaceae(HI)\Pedobacter(HI)\...
982 Unc.(HI)\Sphingopyxis(HI)\...

(continued)

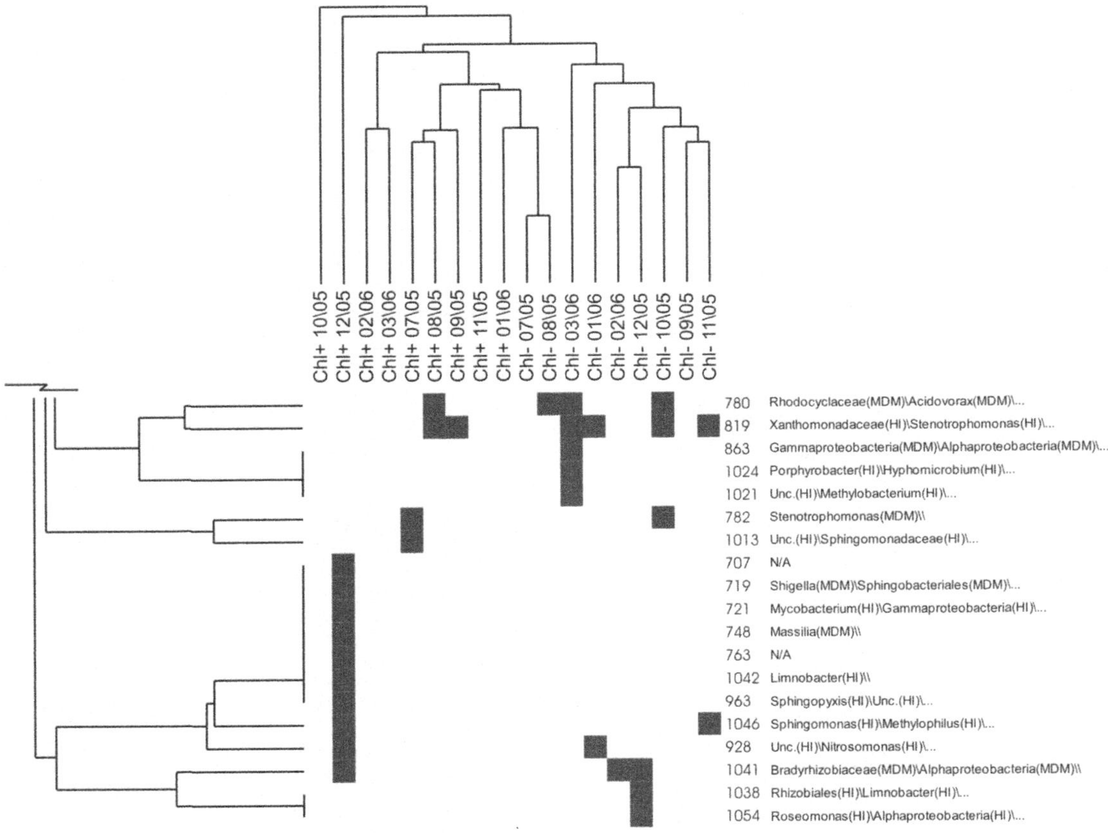

Figure 5.4 Heat-graph representation of hierarchical clustering with bench-scale chemostats. Miniaturized text on the right shows up to two highest confidence matches to the alleles taken from Table A1. Ending with "\\" indicates all assignments are covered; '\.." denotes more assignments are present in Table A1. Script in parentheses indicate the level of confidence in the clone assignments to the allele: HI, high; MDM, medium; LOW, low. Unc., unclassified; N/A, not available.

The figure shows that the clustering of samples according to the similarities on allele patterns resulted in a general separation of Chl- samples from Chl+ samples, although July and August 2005 samples from the Chl- reactor grouped with the January 2006 sample from the Chl+ reactor. October and December 2005 samples of Chl+ were markedly different from the rest of the samples and did not cluster with the other samples. This period of operation corresponded to a transition on the performance of Chl+, evidenced by a decrease in nitrification efficiency and the accumulation of ammonia (Figure 2.2). Towards the end of the operation of both reactors, Chl+ was receiving the highest chloramine dose imposed during the study (greater than 5 $mgCl_2/L$ in the feed on February and March 2006), which appears to be reflected in the community profiles, as evidenced by these two samples forming a well-defined cluster in Figure 5.4.

The vertical clustering grouped alleles mainly based on their patterns of presence/absence in the samples from the two reactors. For instance, the upper nodes in Figure 5.4 corresponded to alleles frequently found in Chl+ samples and only sporadically present in Chl- samples, which could represent organisms that are not negatively impacted by the presence of chloramines. This region includes alleles such as 759, 798, 893, 978, which could be potentially matched to *Mycobacterium,* members of the order *Sphingobateriales*, and *Caulobacter*, respectively, in agreement with the large numbers of sequences representing these groups within Chl+ as compared to Chl- (Tables 4.1 and 4.2). Figure 5.4 also shows nodes where alleles found only in Chl- also grouped together, and may represent organisms that are sensitive to chloramines. Among this group are alleles 912 and 925, which could be linked to *Acinetobacter*, a genus repeatedly found in Chl- but not in Chl+ (Tables 4.1 and 4.2). Further down, Figure 5.4 shows a cluster of alleles present in both reactors. Some of these alleles can be potentially matched to the genera *Nitrospira*, and *Stenotrophomonas*, which were also common to to both reactors in 16S sequences (Tables 4.1 and 4.2).

While a general consistency between allele assignments to 16S identities and the clone libraries in Tables 4.1 and 4.2 was evident, some discrepancies were also observed. For example, allele 628 is present only in Chl- profiles and is associated only with *Achromobacter* (Figure 5.4). In contrast, it was found among Chl+ clones relatively frequently (Table 4.2) but not at all among Chl- clones (Table 4.1). However, given the redundancy of allele matching and the resource limitations for sequencing, one cannot rule out the significant possibility that this allele may also be associated with another organism that is not yet present in the clone library or could not be matched to alleles with reliable clone ARISA profiles.

Pilot-Scale Systems

The clustering of all pilot-scale profiles is depicted in Figure 5.5. Although only a few samples were available from Louisville pilot plants, they clearly formed a separate cluster from UW plants. However, LO1 and LO3 did not form separate clusters, even though LO1 was not chloraminated while LO3 was chloraminated and also received chlorite. Thus, the clustering analysis suggests that the differences in operation were not sufficient to create community shifts between LO1 and LO3.. This observation is consistent with 16S rRNA gene sequences obtained from LO1 (Table 4.5) and LO3 (Table 4.6), which also showed similar community composition, and, as mentioned above, might be explained by the fact that sampling was done in the early phase of LO1 pilot plant (Figure 6.1), possibly not allowing significant community divergence to take place.

The clustering of the two UW plants showed that an important parameter was the period of sampling. The first sampling period was from July to November 2005 and the second one was from August 2006 to January 2007 (Figure 5.1). All UWL samples from the 2005 formed a separate cluster, while most of the UWL samples from the second period also grouped together, with the exception of the sample taken on September 2006. Likewise, most to the UWH samples grouped according to their sampling period. Two samples, UWH July 2005 and UWL September 2006, were significantly distant from the rest of the profiles.

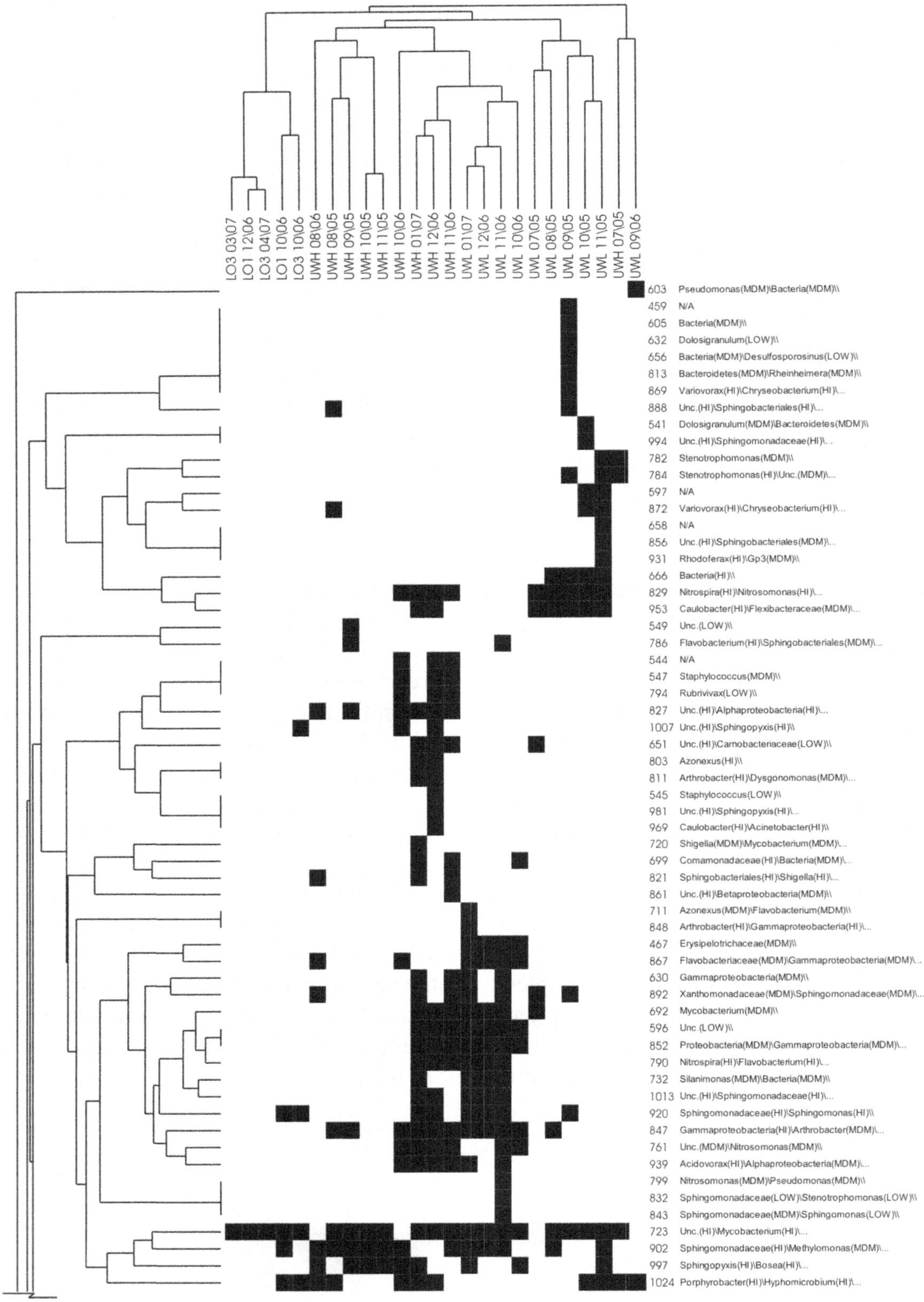

LO3 03\07
LO1 12\06
LO3 04\07
LO1 10\06
LO3 10\06
UWH 08\06
UWH 08\05
UWH 09\05
UWH 10\05
UWH 11\05
UWH 10\06
UWH 01\07
UWH 12\06
UWH 11\06
UWL 01\07
UWL 12\06
UWL 11\06
UWL 10\06
UWL 07\05
UWL 08\05
UWL 09\05
UWL 10\05
UWL 11\05
UWH 07\05
UWL 09\06
603 Pseudomonas(MDM)\Bacteria(MDM)\\
459 N/A
605 Bacteria(MDM)\\
632 Dolosigranulum(LOW)\\
656 Bacteria(MDM)\Desulfosporosinus(LOW)\\
813 Bacteroidetes(MDM)\Rheinheimera(MDM)\\
869 Variovorax(HI)\Chryseobacterium(HI)\...
888 Unc.(HI)\Sphingobacteriales(HI)\...
541 Dolosigranulum(MDM)\Bacteroidetes(MDM)\\
994 Unc.(HI)\Sphingomonadaceae(HI)\...
782 Stenotrophomonas(MDM)\\
784 Stenotrophomonas(HI)\Unc.(MDM)\...
597 N/A
872 Variovorax(HI)\Chryseobacterium(HI)\...
658 N/A
856 Unc.(HI)\Sphingobacteriales(MDM)\...
931 Rhodoferax(HI)\Gp3(MDM)\\
666 Bacteria(HI)\\
829 Nitrospira(HI)\Nitrosomonas(HI)\...
953 Caulobacter(HI)\Flexibacteraceae(MDM)\...
549 Unc.(LOW)\\
786 Flavobacterium(HI)\Sphingobacteriales(MDM)\...
544 N/A
547 Staphylococcus(MDM)\\
794 Rubrivivax(LOW)\\
827 Unc.(HI)\Alphaproteobacteria(HI)\...
1007 Unc.(HI)\Sphingopyxis(HI)\\
651 Unc.(HI)\Carnobacteriaceae(LOW)\\
803 Azonexus(HI)\\
811 Arthrobacter(HI)\Dysgonomonas(MDM)\...
545 Staphylococcus(LOW)\\
981 Unc.(HI)\Sphingopyxis(HI)\...
969 Caulobacter(HI)\Acinetobacter(HI)\\
720 Shigella(MDM)\Mycobacterium(MDM)\...
699 Comamonadaceae(HI)\Bacteria(MDM)\...
821 Sphingobacteriales(HI)\Shigella(HI)\...
861 Unc.(HI)\Betaproteobacteria(MDM)\\
711 Azonexus(MDM)\Flavobacterium(MDM)\\
848 Arthrobacter(HI)\Gammaproteobacteria(HI)\...
467 Erysipelotrichaceae(MDM)\\
867 Flavobacteriaceae(MDM)\Gammaproteobacteria(MDM)\...
630 Gammaproteobacteria(MDM)\\
892 Xanthomonadaceae(MDM)\Sphingomonadaceae(MDM)\...
692 Mycobacterium(MDM)\\
596 Unc.(LOW)\\
852 Proteobacteria(MDM)\Gammaproteobacteria(MDM)\...
790 Nitrospira(HI)\Flavobacterium(HI)\...
732 Silanimonas(MDM)\Bacteria(MDM)\\
1013 Unc.(HI)\Sphingomonadaceae(HI)\...
920 Sphingomonadaceae(HI)\Sphingomonas(HI)\\
847 Gammaproteobacteria(HI)\Arthrobacter(MDM)\...
761 Unc.(MDM)\Nitrosomonas(MDM)\\
939 Acidovorax(HI)\Alphaproteobacteria(MDM)\...
799 Nitrosomonas(MDM)\Pseudomonas(MDM)\\
832 Sphingomonadaceae(LOW)\Stenotrophomonas(LOW)\\
843 Sphingomonadaceae(MDM)\Sphingomonas(LOW)\\
723 Unc.(HI)\Mycobacterium(HI)\...
902 Sphingomonadaceae(HI)\Methylomonas(MDM)\...
997 Sphingopyxis(HI)\Bosea(HI)\...
1024 Porphyrobacter(HI)\Hyphomicrobium(HI)\...

(continued)

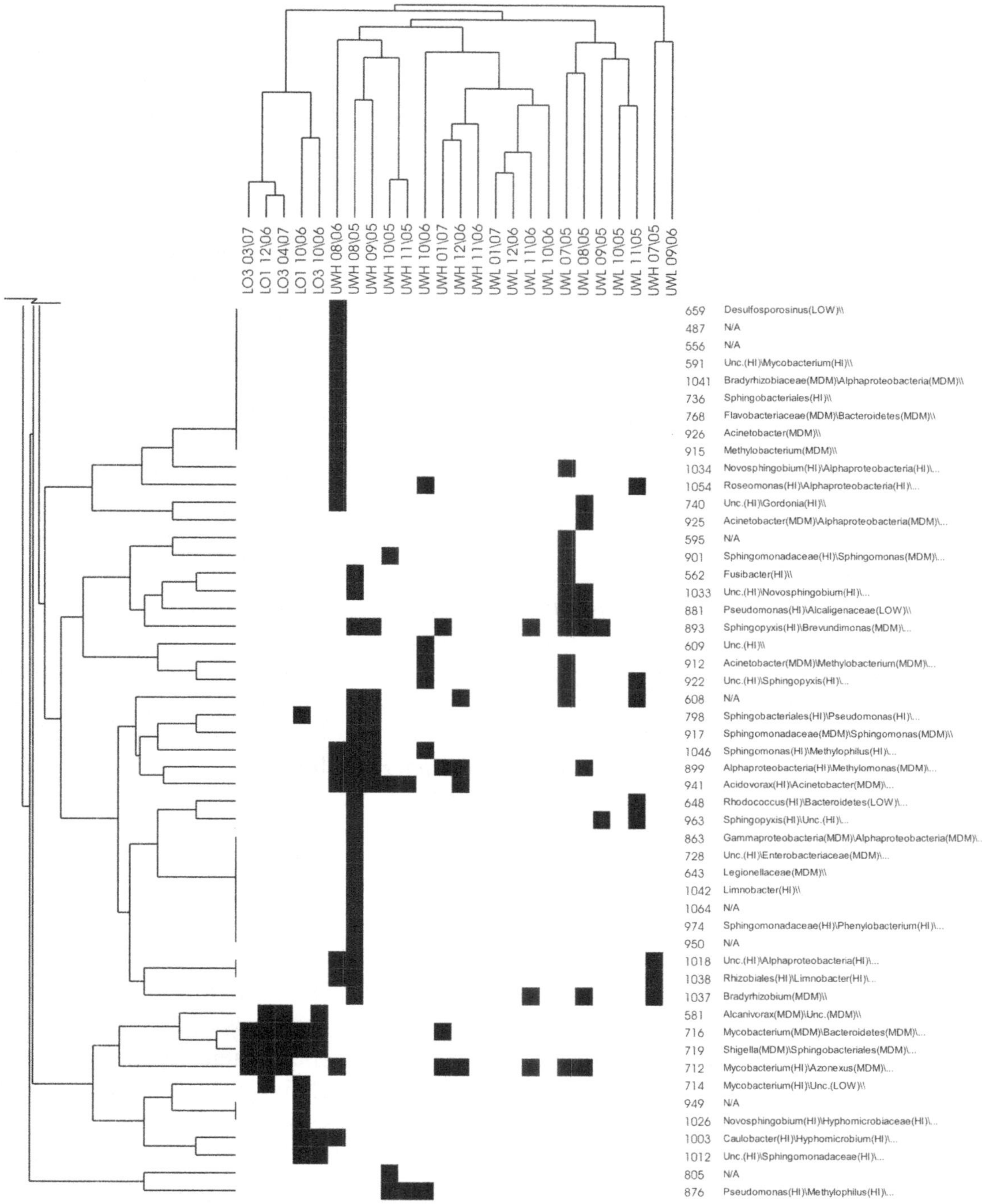

Figure 5.5 Heat-graph representation of hierarchical clustering with pilot-scale distribution systems. Miniaturized text on the right shows up to two highest confidence matches to the alleles taken from Table A1. Ending with "\\" indicates all assignments are covered; '\.." denotes more assignments are present in Table A1. Script in parentheses indicate the level of confidence in the clone assignments to the allele: HI, high; MDM, medium; LOW, low. Unc., unclassified; N/A, not available.

Reflecting the lack of good separation between UWH and UWL profiles, alleles were generally clustered into groups appearing in both distribution systems (Figure 5.4). However, a cluster of generally UWH-associated peaks was apparent in the middle of the first half of Figure 5.4. Interestingly, this cluster included two alleles, 545 and 547, matching only to *Staphylococcus*, which was not found in any of the pilot-scale clone libraries (Tables 4.3 and 4.4). Similarly, peak 803 matched only to *Azonexus*, also not found among pilot-scale clones. Once again, such discrepancies might be due to incomplete assignment of clones to fragment lengths as well as to the insufficient sampling of clones. Another UWH cluster is located at the bottom of Figure 5.4. This group could be associated with *Pseudomonas*, which was present in large numbers in the UWH clone library (Table 4.4), as opposed to UWL (Table 4.3), and created the biggest difference between the two libraries.

An interesting cluster towards the bottom of the first half of Figure 5.4 consisted mostly of 2006-2007 run samples from both UWH and UWL systems. The associated alleles included common organisms such as *Mycobacterium* (allele 692) and members of the *Sphingomonadaceae* family within the *Alphaproteobacteria* (alleles 920 and 1013), according to clone-ARISA assignments. Another interesting branch at the bottom of the first half of Figure 5.4 comprised alleles observed in all systems including Louisville. Consistent with clone libraries (Tables 4.3, 4.4, 4.5, and 4.6) these alleles could be matched to the commonly found *Mycobacterium*, *Sphingomonas*, *Hypobacterium*, and *Porphyrobacter* genera. Most of the remaining alleles associated with the Louisville system formed a separate cluster towards the end of Figure 5.4, where clone ARISA matches indicated *Mycobacterium* and *Sphingomonas* as probable organisms, also consistent with clone libraries (Tables 4.5 and 4.6). Finally, alleles potentially related to two phylogenetically distant but phenotypically related groups of bacteria, namely the *Sphingomonadales* order of the *Alphaproteobacteria* (alleles 893, 901, 963) and the *Sphingobacteriales* order of the *Bacteroidetes* (alleles 821, 856) were frequent in the remaining parts of the allele tree, where distribution of peaks to samples generally showed patchy patterns, emphasizing the commonality of these lineages in the pilot-scale distribution systems.

Full-Scale Systems

The clustering of full-scale samples is depicted in Figure 5.6. Unlike bench- and pilot-scale samples, no clear subclusters of individual sampled locations were observed. As such, the microbial communities in the Martin's Hill Tank and and the Prospect Tank appeared to have significant overlap. Morever, since four out of the seven samples of finished water collected at the treatment plant did not yield sufficient DNA for ARISA analyses, it is difficult to conclude on the relationship of these samples to those collected at the other two locations, although one of the finished samples (November 2006) exhibited a large number of peaks not found in any of the other samples.

Consistent with the mixed clusters of samples, alleles were sporadically distributed to different locations, and therefore, distinct clusters of a particular location were not very notable. Exceptions included a small cluster of alleles found only in samples from the Prospect Tank (888, 1021, and 1012), which could potentially be associated with *Sphingomonas*- and *Methylobacterium*-related organisms also found in the 16S rRNA gene sequences from this tank (Table 4.8).

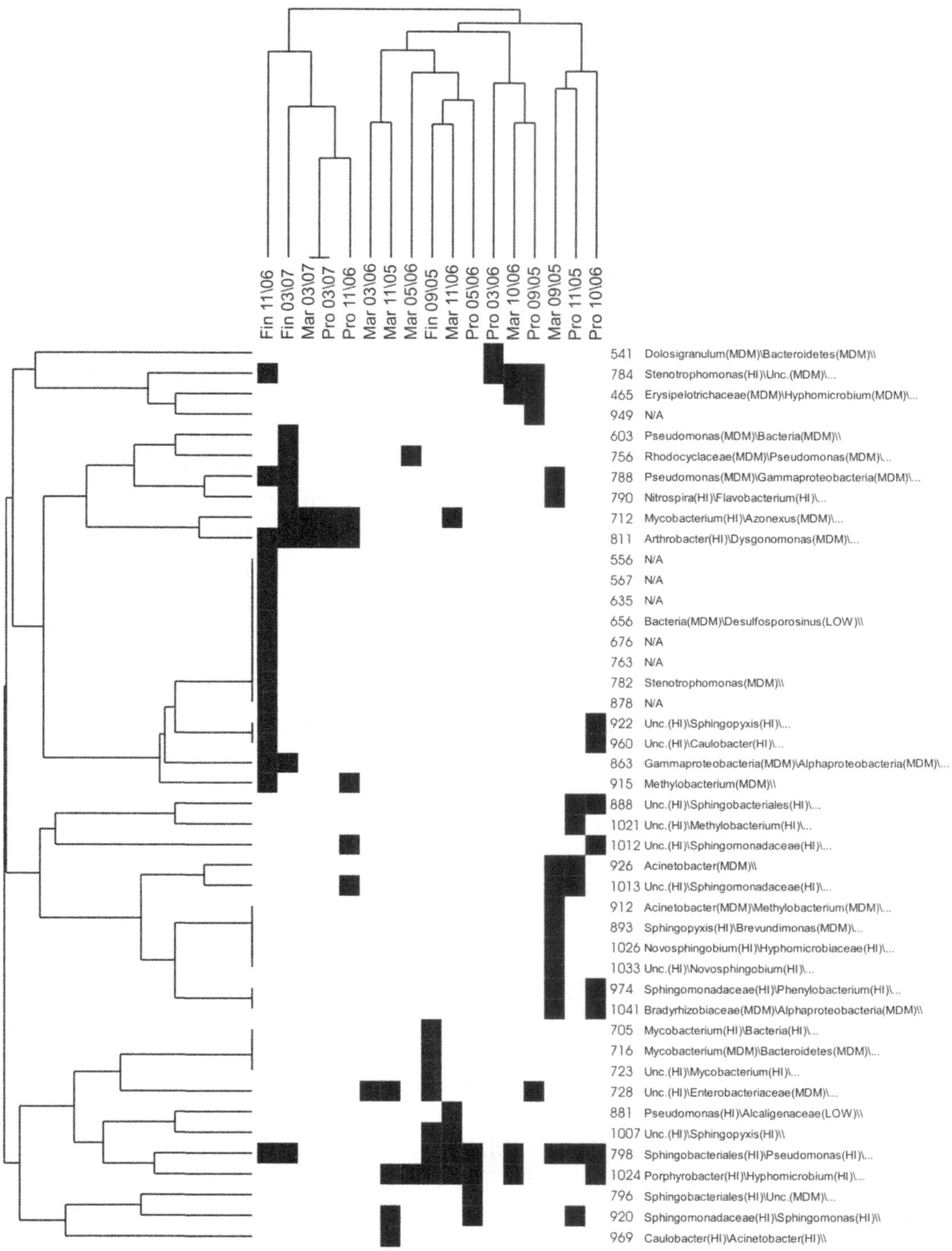

Figure 5.6 Heat-graph representation of hierarchical clustering with full-scale distribution systems. Miniaturized text on the right shows up to two highest confidence matches to the alleles taken from Table A1. Ending with "\\" indicates all assignments are covered; '\.." denotes more assignments are present in Table A1. Script in parentheses indicate the level of confidence in the clone assignments to the allele: HI, high; MDM, medium; LOW, low. Unc., unclassified; N/A, not available.

Another exception was a dense branch of alleles found only in one finished water sample from November 2006 (556, 567, 635, 656, 676, 763, 782, and 878). One of these peaks could be associated with *Stenotrophomonas*, which agrees with the finished water clone library, since two *Xanthomonadacea* sequences were not classified at the genus level (Table 4.7) and can potentially be representing *Stenotrophomonas*. However, *Stenotrophomonas*-related organisms were also found in the Martin's Hill Tank sequences (Table 4.9). More importantly, most of these finished water peaks could not be matched to any clones in the overall library (Figure 5.6). This could be due to the fact that all finished water cloning/sequencing was done using two passes of PCR, potentially resulting in a different community than in one-pass ARISA profiles, a methodology question that requires further investigation.

A small cluster of two alleles (798 and 1024) towards the bottom of Figure 5.6 were present in all three locations. The allele 798 was potentially associated with the *Sphingobateriales* order within *Bacteroidetes* and with *Pseudomonas*-related bacteria commonly found in all three locations (Tables 4.7, 4.8, and 4.9), while 1024 matched to *Hyphomicrobium* found in finished water clone library (Table 4.7) and *Methylobacterium* (Table A1) found in the Prospect Tank (Table 4.8) and the Martin's Hill Tank (Table 4.9) sequences. Finally, a general observation from Figure 5.6 was the frequent appearance of alleles potentially related to *Mycobacterium* (e.g. 712, 705, 716, and 723) and the *Sphingomonadales* and *Sphingobateriales* orders (e.g. 922, 888, 1012, 1013, 893, 1026, 1033, 1007, 798, 926, and 920) over the vertical axis, agreeing with the widespread occurrence of these groups in clone libraries and other ARISA profiles.

NON-METRIC MULTIDIMENSIONAL SCALING (MDS) ANALYSIS OF ARISA PROFILES

Non-metric multidimensional scaling (MDS) has been proposed as a powerful method for exploring patterns in ARISA fingerprints that cannot be captured visually (Yannarell and Triplett 2004). Here, it was employed for the presentation of the overall clustering of samples profiled, and for subsequent statistical analyses of the strength and significance of separations between the different distribution systems studied.

As with hierarchical clustering methods, the profiles were first transformed to absence/presence vectors and a similarity matrix was generated for the quantification of distance between these vectors. Then, clustering analysis was carried out using the similarity matrix. All steps were performed using the Primer 6.0 software as explained in Chapter 2.

The visual output of the analysis, shown in Figure 5.7, is a 2D projection of data points representing ARISA profiles. In this type of MDS plot, similar profiles are positioned proximately, while distant ones are separated apart. The concept of similarity-based positioning is illustrated in Figure 5.7 using the data points that represent ARISA profiles of Chl+, Chl-, and Prospect samples shown in Figure 5.2.

An important observation from Figure 5.7 was the distinct clusters of systems from different scales. In addition to the clustering plots, Primer 6.0 software was used also for statistically testing the significance of such observations. To this end, the profiles were grouped first into three different scales, and then into different distribution systems. The strength of separation between the clusters of different groups was measured based on Spearman's rank correlation (R; the closer R is to 1 the stronger the separation between two given groups), which was associated with a p-value indicating the statistical significance of the separation. A p-value of 0.05, which shows significant separation at the 95% confidence level, was taken as the threshold for calling the separation of a pair statistically significant. Table 5.1 lists the results of statistical tests. Thus, the presence of distinctive clusters representing different scales was confirmed at 0.1% level of significance ($p < 0.001$). Consistent with visual inspection of Figure 5.7, the relatively high *R*-values for bench-pilot ($R = 0.79$) and bench-full (0.69) pairs indicate a strong separation of bench-scale systems from the others, while the lower *R*-value for pilot-full pair (0.47) reflects the observed overlap between the pilot- and bench-scale community profiles. As discussed in Chapter 4, while the differences may be due in part to system scale, it is important to recognize that the chemical water quality differences described in Chapter 2 may also have a significant influence on these results.

Pairwise statistical tests at the second (systems) level, also seen in Table 5.1, were generally in agreement with the clustering patterns of samples in Figures 5.4, 5.5, and 5.6. Thus, Chl+ and Chl- showed relatively weak ($R = 0.3$) but statistically significant ($p < 0.001$) separation, pointing to the significant effect of chloramination as a driver of community composition. However, UWH and UWL systems did not form separate patterns ($p >> 0.05$), suggesting that the targeted level of differentiation in the disinfection level did not cause major changes in bacterial community in these systems. On the other hand, the Louisville pilot plants formed a clustered together ($R = 0$), which was distinct from UWH ($p < 0.03$), but not from UWL ($p > 0.05$), although the tests on pairs involving Louisville plants should be interpreted with care due to the low number of samples from these plants and to the aforementioned possibility that a true non-chloraminated system community may not have been established in the LO1 plant during the early phase of the run when the sampling took place (see Figure 5.1). As for the full-scale distribution system locations, statistical tests suggested that none of the pairs were significantly different from each other, as indicated by negative *R*-values and large p-values (Table 5.1). Finally, pairs across different scales were significantly separated as expected from the different clusters of scales, except for the Louisville pilot-plants not significantly different from full-scale locations. However, once again, this was likely because of the low number of samples from Louisville plants, but could also point out to the impact of source water on the community composition within a drinking water distribution system.

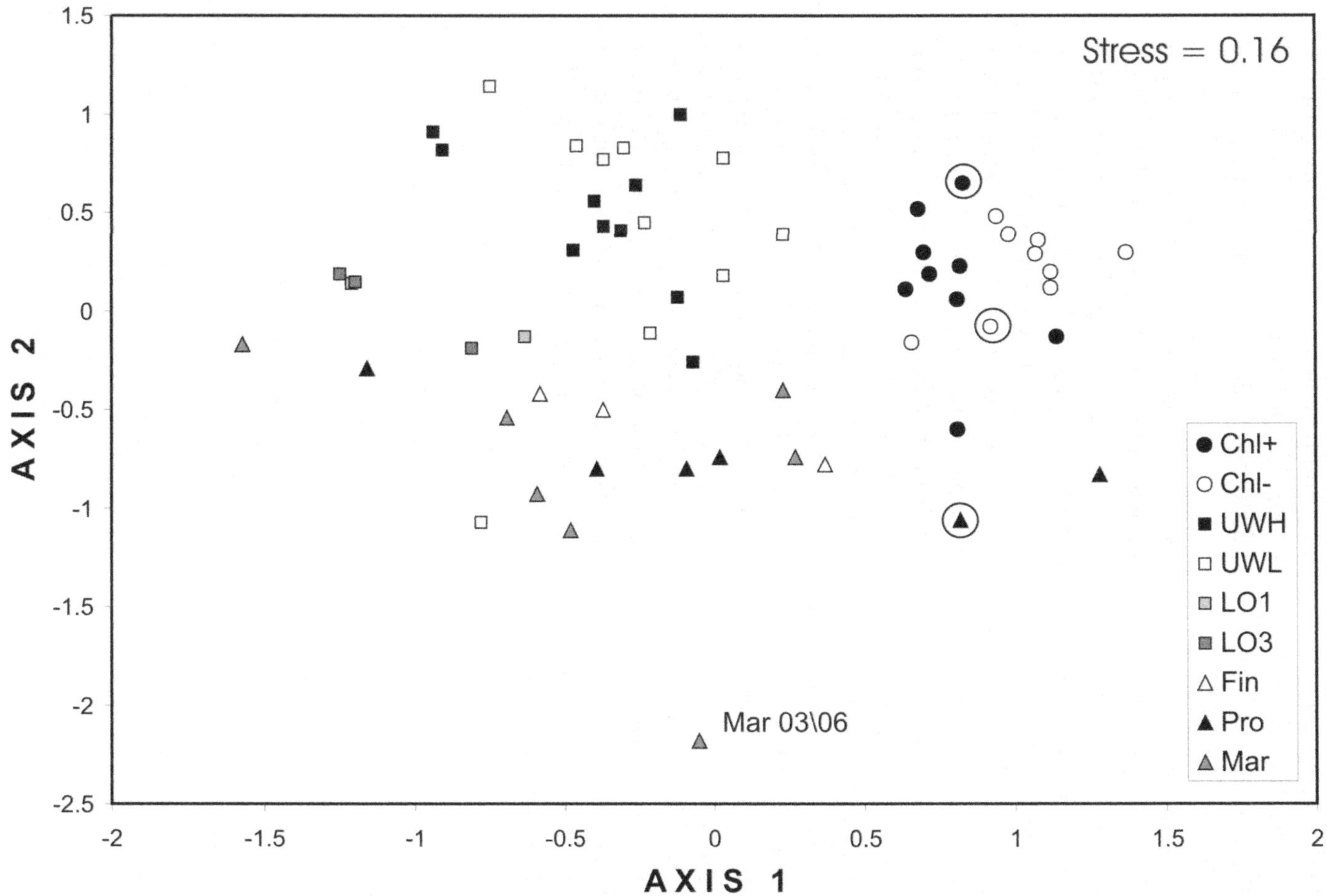

Figure 5.7 MDS plot showing the clustering of all community ARISA profiles. Data points that cover the samples in Figure 5.2 are circled to aid in visualization. Martin sample from March 2006 is labeled since it appears as an outlier. Stress indicates the level of compromise due to the projection onto 2D space, as calculated from Kruskal stress formula (Kruskal 1964).

Table 5.1
Statistical analysis of clustering patterns in Figure 5.7*

	Bench	Pilot	Full	Chl+	Chl-	UWH	UWL	LO1	LO3	Fin	Pro	Mar
Bench		0.79 (<0.001)	0.63 (<0.001)									
Pilot			0.47 (<0.001)									
Full												
Chl+					0.30 (<0.003)	0.90 (<0.001)	0.70 (<0.001)	1.00 (<0.018)	1.00 (<0.005)	0.9 (<0.005)	0.58 (<0.001)	0.68 (<0.001)
Chl-						0.95 (<0.001)	0.73 (<0.001)	1.00 (<0.018)	1.00 (<0.005)	0.98 (<0.005)	0.65 (<0.001)	0.74 (<0.001)
UWH							0.04 (<0.212)	0.45 (<0.03)	0.60 (<0.003)	0.69 (<0.007)	0.60 (<0.001)	0.61 (<0.001)
UWL								0.23 (<0.182)	0.30 (<0.056)	0.50 (<0.014)	0.41 (<0.001)	0.42 (<0.002)
LO1									0.00 (<0.600)	0.33 (<0.2)	-0.04 (<0.583)	0.12 (<0.278)
LO3										0.50 (<0.1)	0.13 (<0.183)	0.24 (<0.125)
Fin											-0.23 (<0.958)	-0.15 (<0.842)
Pro												-0.08 (<0.749)
Mar												

*Numbers indicate the R-statistic (outside parentheses) and p-statistic (inside parentheses).

CHAPTER 6
RESULTS AND DISCUSSION: CAN THE MOLECULAR ECOLOGICAL DATABASE BE USED FOR THE IDENTIFICATION OF ORGANISMS IN OTHER DISTRIBUTION SYSTEMS?

The sequencing and clone-ARISA fingerprinting data in Chapters 4 and 5 formed a large database of about 1,000 sequenced clones associated with certain lengths of ITS-based fragments. In this chapter, the possibility of using the clone-ARISA database for predicting the identity of organisms found in other drinking water distribution systems was investigated. Samples were collected from two municipal (or community water supply) distribution systems, denoted here as Utility X and Utility Y.

The underlying idea is the potential identification of distribution system organisms based on ARISA fingerprints, which are relatively easy to obtain as compared to 16S rRNA sequences. Because mixed community ARISA profiles do not allow direct identification of peaks for evaluating the predictions, studied drinking water samples needed to be cultured for a direct linkage of ARISA alleles to 16S-based identity. The obtained ITS-16S matches could then be compared to those in the clone-ARISA database developed. This culturing requirement is consistent with practices commonly used at utilities for monitoring microbial growth in distribution systems.

EXPERIMENTAL PROCEDURE

Due to the culture dependent nature of the approach, the experimental methods used in this task were slightly different from the general procedure explained in Chapter 2. For Utility X, samples from the distribution system were received from the utility as plates showing colony growth. We received plates prepared in Standard Plate Count Agar (pour plate method, incubated at 35 °C for 48 hrs) and Sheep Blood Agar (290 mL water sample, membrane filtered, incubated at 35 °C for 24 hrs). For Utility Y, spread plates were prepared on R2A agar and incubated for at least 5-day at room temperature. Visible colonies on plates were transferred to 96-well plates in preparation for ARISA analyses (Figure 6.1). After releasing a colony on a well containing PCR buffer, cells were dispersed and re-suspended by passing the liquid multiple times through a narrow bore needle. Using the cell suspensions as templates, ARISA PCR reactions were carried out as explained before with labeled 1406F primer and unlabeled 23Sr primer. ARISA profiles were generated as with other drinking water samples (see Chapter 3).

To associate ARISA peaks with the actual identity of the organisms, colony suspensions were sequenced by first amplifying the 16S gene with 27F and 1492R primers in separate PCR reactions, and then partial sequencing of the 16S rRNA with 27F as described before (see Chapter 3).

In the ideal case, the applied procedure gives a single ARISA peak for a colony representing a pure culture, which can then be matched to a predicted identity in the database of alleles obtained from clone-ARISA (Figure 6.1). The accuracy of the prediction can then be tested based on the true identity from the 16S rRNA sequence of the pure culture. An example single-peak ARISA profile is given in Figure 6.2a. However, in many cases, multiple ARISA peaks were observed (Figure 6.2b). This was anticipated since multiple copy numbers of the rRNA operon is common among prokaryotes (Klappenbach et al. 2001). Although multiple alleles could also imply that the suspended colony was a mixed culture, it was assumed to be a pure culture if a high quality sequence was achieved from the same template. Thus, identities in clone-ARISA database matching to all observed ARISA peaks of a given colony were used as predictions.

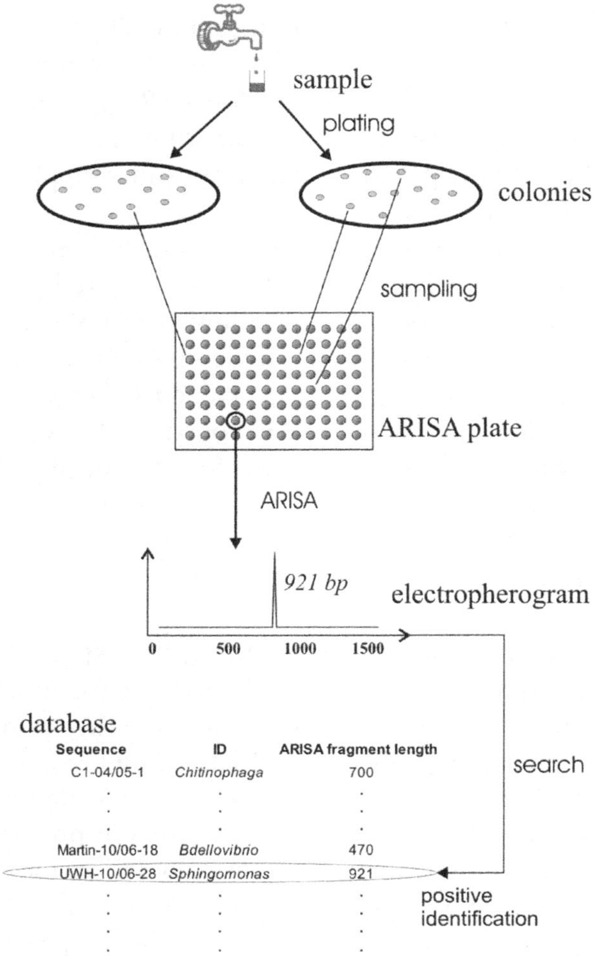

Figure 6.1 Procedure for the identification of colonies obtained from drinking water samples. bp, base pairs.

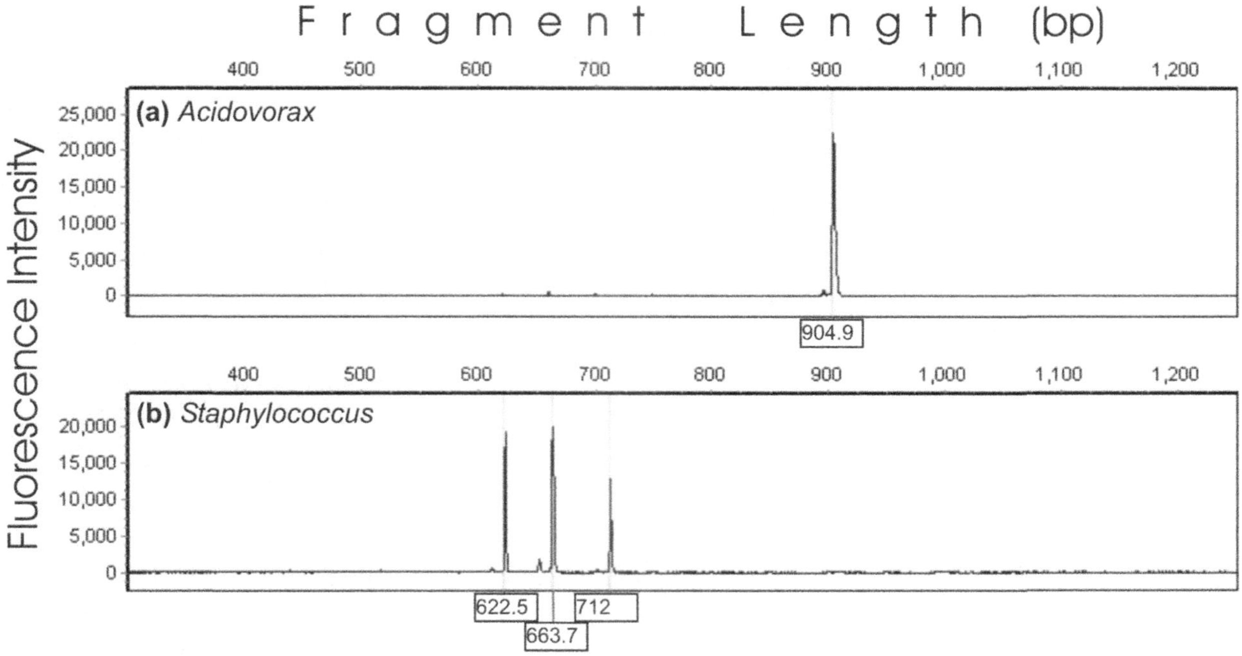

Figure 6.2 Selected ARISA profiles obtained from drinking water colonies exemplifying organisms associated with a single peak (a) and multiple peaks (b). Genus-level identities of organisms are indicated based on matching 16S rRNA sequences, according to the RDP classifier tool.

IDENTIFICATION OF COLONIES

In a total of 16 colonies that were successfully analyzed for both ARISA profiles and 16S rRNA gene sequences, three genotypes were observed for source X and one for source Y (Table 6.1). Only one genotype was associated with a single allele. While 16S sequencing showed that this organism was a member of the *Acidovorax* genus within the *Betaproteobacteria*, the clone-ARISA database matched the allele to *Methylophilus* (*Betaproteobacteria*), *Flavobacterium* (*Bacteroidetes* phylum), and *Legionella* (*Gammaproteobacteria*). For other organisms, multiple peaks were observed and a large set of matches was generally found in the database. One of the peaks of a colony identified as *Blastomonas*, a genus related to *Sphingomonas*, was correctly predicted, although other matches of the same allele masked this prediction. In another case, one of the six peaks of *Paenibacillus* (*Firmicutes* phylum) was linked to a clone that was identified as *Firmicutes* at the phylum level with >95% confidence (Table 6.1), as *Bacillales* with a relatively low but considerable 89% confidence (not shown in Table 6.1), which placed this clone in the same family with *Paenibacillus*, but could not be further classified with reasonable confidence. Finally, a *Staphylococcus*-related colony (*Firmicutes* phylum) from source X could not be matched to any relevant clones in the database, although it was associated with a variety of other organisms including *Mycobacterium* (*Actinobacteria* phylum) and *Enterobacteriaceae* (*Gammaproteobacteria*) from different phyla than *Staphylococcus*.

TABLE 6.1
List of identified colonies

Organism	Source	Allele	Number of colonies	Colony identification*	Predicted Identification from ARISA peak†
1	X	905	3	*Acidovorax*	*Methylophilus* (HI) \ *Flavobacterium* (MDM) \ *Legionella* (MDM) \
2	X	997	5	*Blastomonas*	*Sphingopyxis* (HI) \ *Bosea* (HI) \ *Alphaproteobacteria* (HI) \ *Hyphomicrobiaceae* (MDM) \
		1003	5	*Blastomonas*	*Hypomycrobium* (MDM) \ *Blastomonas* (MDM)
3	X	623	1	*Staphylococcus*	*Enterobacteriaceae* (HI) \ *Rhodococcus* (LOW) \
		664	1	*Staphylococcus*	*Mycobacterium* (HI) \ *Enterobacteriaceae* (MDM) \
		712	1	*Staphylococcus*	*Mycobacterium* (HI) \ *Azonexus* (MDM) \ Unc. (MDM) \ *Flavobacterium* (MDM) \
4	Y	596	7	*Paenibacillus*	N/A
		598	7	*Paenibacillus*	N/A
		621	7	*Paenibacillus*	*Spirochaetaceae* (MDM) \
		668	7	*Paenibacillus*	*Firmicutes* (MDM) \
		915	7	*Paenibacillus*	*Methylobacterium* (MDM) \
		917	7	*Paenibacillus*	*Sphingomonadaceae* (MDM) \ *Sphingomonas* (MDM) \

*Identity of 16S sequence based on a 95% confidence in RDP Classifier (Cole et al. 2005).
†Identity of clones (also based on a 95% confidence in RDP Classifier) with matching clone-ARISA peaks. Script in parentheses indicate the level of confidence in the matching (HI, high confidence; MDM, medium; and LOW, low).

CHAPTER 7
CONCLUSIONS AND RECOMMENDATIONS

CONCLUSIONS

While nitrifiers inhabiting chloraminated distribution systems are relatively well studied (Regan et al. 2002; Regan et al. 2003), this report aimed at the identification of the heterotrophic component of the microbial population in bulk water from such systems. An overall general observation was that the 16S rRNA gene sequences and the community fingerprints obtained with universal primers were dominated by heterotrophic bacteria, while nitrifiers formed a minority. Thus, the original aim of focusing only on heterotrophic bacteria was fulfilled without a need for specific methods to exclude the nitrifier population. Although PCR methods introduce a bias that prevents a quantitative conclusion of microorganism abundance based on the frequency of sequences observed, the fact that a diverse and large set of organisms was retrieved suggests that heterotrophic bacteria dominated the communities in the different chloraminated environments studied. The dominance of heterotrophic bacteria over nitrifiers in nitrifying environments is commonly assumed (AWWA 2006) and will be typical of environments where sufficient biodegradable organic matter is available, but an ultimate demonstration that this is indeed the case in chloraminated drinking water distribution systems would require a purposefully directed quantification using culture independent methods such as fluorescence in situ hybridization (Wagner et al. 2003).

At the phylum level, the library of 16S rRNA gene sequences obtained from full-scale distribution system samples collected by the Louisville Water Company was mainly composed of members of the phyla *Actinobacteria*, *Bacteriodetes*, *Firmicutes*, and *Proteobacteria* at every location sampled. The three locations showed significant similarities also at lower taxonomic levels, as evident from family and genus level classifications of clone libraries and ARISA profiles. For example, *Azonexus*-related species were common to all locations including finished water. Therefore, the community composition in the distribution system was likely affected by the source of water and the input from the treatment plant. Support for this is also evident by similarities observed between the communities in the full-scale samples and those found in the pilot-scale systems operated at Louisville. However, differences between finished water and downstream locations were also evident. For example, *Alcanivorax*-related sequences were retrieved only from finished water samples, while *Sphingomonas*-related species (members of the order *Sphingomonadales*) were found only in downstream locations, and the frequency of *Mycobacterium*-related clones significantly increased when going from finished water to distribution system tanks. Also, potential nitrifiers such as members of the *Nitrosomonas* and *Nitrospira* lineages appeared only in distribution system tanks, as expected. Thus, the expected diversification of community from the seed of the distribution system, potentially due to interactive nitrification and heterotrophic growth, was also supported by the results. On the other hand, the community profiles of the two distribution system tanks studied (Prospect and Martin's Hill tanks), seemed quite similar based on both 16S rRNA and ITS signatures. This result could be explained if one assumes that community profile is a function of the seed from the treated water and environmental conditions governed by the disinfection strategy and geographical area.

A particular objective of this study was to evaluate whether chloraminated bench- and pilot-scale systems could represent full-scale environments with regard to bacterial community composition. ARISA fingerprints showed statistically significant differences between the scales, with some overlap between pilot- and full-scale. These differences in scale may be due to differences in the chemical composition of the water from one source to another. For example, no significant differences could be detected between the pilot-scale and full-scale samples at Louisville. While these results pointed to differentiation at species level (Brown et al. 2005), 16S sequences and their clustering in phylogenetic trees indicated remarkable similarities at higher taxonomic levels. Thus, controlled reactors and simulators, as those sampled in this study, can be used to grow and investigate types of organisms inhabiting real scale distribution systems, but representative ability at the species level should be handled with care. It was also evident in this study that bench-scale reactors were more prone to contamination by other organisms present in the laboratory, which makes them less ideal for meaningful community composition studies.

A striking similarity between different scales was the presence of organisms related to the *Mycobacterium* genus within the *Actinobacteria* phylum, and the *Sphingomonadales* order within *Alphaproteobacteria* in all chloraminated systems studied. Interestingly, the nitrifying bench-scale reactor that did not receive any chloramines was short of these types of organisms, while its chloraminated parallel yielded a large set of clones for each type. Consistently, sequences associated with *Mycobacterium* and *Sphingomonadales* were not found frequently in finished water samples, while they were dominant in downstream locations. These results provide a direct connection between the prevalence of these organisms and chloramination.

A large set of OTUs from all levels of scales formed noticeable clusters with public sequences also retrieved from drinking water environments (data available in phylogenetic trees of Chapter 4). Most, but not all, of these drinking water clades involved chloraminated systems (Williams et al. 2004, and Williams et al., unpublished). Yet, some other sequences obtained in this study formed previously not detected clusters. While the significant overlaps and similarities with previous studies confirmed the importance of residual chloramines and other factors associated with distribution systems as drivers of community composition, novel organisms contributed significantly to the expanding database of drinking water microbial ecology. About 1,500 partial sequences generated in this study more than doubled the information on drinking water associated bacteria in the comprehensive database of RDP Release 9 (Cole et al. 2005). Around 200 of these were converted to full sequences as representatives of selected OTUs, which amounted to approximately half of the full drinking water sequences publically available in the same database. Nevertheless, the 16S sequence library generated in this project should still be considered as providing a brief snapshop of the heterotrophic microbial community that may be present in chloraminated distribution systems. It is likely that many site-specific issues, such as water age, pipe materials, temperature, and operational practices will also affect the composition of the microbial communities. In addition, the absence of biofilm samples in this study is an important gap that should be evaluated in further studies.

The overall knowledge on the molecular ecology of distribution system bulk water can be potentially useful for understanding the factors affecting water quality as related to biological activity. The significant overlaps of different systems point to the presence of a "normal" distribution system microbiota. Because of the relatively large amount of information available from this study and Williams et al.'s work (Williams et al. 2004, and Williams et al., unpublished), a normal microbiota is more evident for chloraminated systems as a separate habitat, which have a large intersection set of 16S-based identities, including *Mycobacterium*- and *Sphingomonas*-related organisms. On the other hand, the differences within chloraminated systems and between these systems and others not only denote diversity, but may also be useful in understanding the drivers of community composition and interactive relationships between different types of organisms such as nitrifiers and heterotrophs.

An additional molecular level database generated in this study was the library of ARISA peaks for most of the recovered clones, which represented the variability in ITS fragment lengths. Initially, the clone-ARISA database was expected to be useful in tracking different types of organisms in community profiles without need for additional sequencing. However, the predictions of peak identities were masked by the multiplicity of matches to the same allele. Apparently, the large diversity of species, combined with the copy numbers of rRNA operons (Klappenbach et al. 2001), exhausted the potential variability of ITS length. Nonetheless, ARISA fingerprints were still useful for comparative statistical analysis of community profiles, and the probable matching of alleles in these fingerprints to 16S identity generally confirmed the findings from clone libraries.

Finally, clones retrieved from the model and real distribution systems in this study generally were not associated with commensal, recognized, or emerging pathogens of concern in drinking water environments (WHO 2003; Berry et al. 2006). Exceptional species-level associations with known pathogens included 16S clusters around *E. coli*, *Stenotrophomonas maltophilia*, and *Staphylococcus acnes*. Therefore, consistent with the original scope and aims of this study, the relevance of the findings to water quality should be directed more towards biostability of distribution systems than to public health, as the methods used do not allow a proper and unequivocal identification of bacterial pathogens.

RECOMMENDATIONS

For utilities interested in exploring the role of heterotrophs in the onset of nitrification in chloraminated systems, it is recommended that they develop a semi-quantitative understanding of whether nitrifiers are the first colonizers in their distribution system or whether chloramine-resistant heterotrophs initiate water deterioration that results in nitrification. For this, general measurements of heterotrophic and nitrifier growth, such as heterotrophic plate counts and most probable number determinations could be useful to assess the relative growth of these two microbial groups within the distribution system.

For utilities interested in identifying the type of heterotrophic bacteria growing in the distribution systems, cloning and sequencing the 16S rRNA gene is the most direct approach. However, due to the cost of these analyses, the information provided by 16S sequencing is finite and dependent on the number of sequences afforded, and therefore, it is recommended that other methods be used for community profiling along with 16S sequencing. In this report, we demonstrate the usefulness of ARISA as a community fingerprinting tool. Moreover, if a utility decides to use 16S rRNA sequencing for bacterial identification, it is recommended the cloning and amplification of not only the 16S rRNA gene, but also of the ITS region. That way, the full benefit of matching ARISA peaks with bacterial identity could be realized.

Finally, the matching of ARISA peaks with 16S sequence information was demonstrated to be useful for the analysis of bacterial communities from different environments. The library of matching information presented in this report is incomplete due to the obvious resource limitations. It is recommended that such library continue to be expanded to include a more distributed set of environments, including samples from full-scale systems with a variety of water sources, treatment schemes, disinfection practices, etc. Only then, could the full benefit of the ARISA peak identification initiated in this project be fully realized.

APPENDIX A

Table A.1
List of ITS alelles with matching clone identities

Allele	Bin Range	Matching Identity*
318	317.1-318	N/A
444	443.2-444	N/A
459	458.9-459.7	N/A
465	464.5-465.5	*Erysipelotrichaceae* (MDM) \ *Hyphomicrobium* (MDM) \ Unc. (LOW) \
467	466-467	*Erysipelotrichaceae* (MDM) \
487	486.9-487.7	N/A
541	539.5-541.6	*Dolosigranulum* (MDM) \ *Bacteroidetes* (MDM) \
544	543.4-544.2	N/A
545	545-545.8	*Staphylococcus* (LOW) \
547	546.3-547.1	*Staphylococcus* (MDM) \
549	548.7-549.8	Unc. (LOW) \
556	554.8-556	N/A
562	561.8-562.8	*Fusibacter* (HI) \
567	566.6-567.4	N/A
581	580.6-582.1	*Alcanivorax* (MDM) \ Unc. (MDM) \
591	591.1-591.9	Unc. (HI) \ *Mycobacterium* (HI) \
595	594.3-595.1	N/A
596	595.2-596.4	Unc. (LOW) \
597	596.5-597.6	N/A
603	601.9-603.1	*Pseudomonas* (MDM) \ Bacteria (MDM) \
605	604.7-605.5	Bacteria (MDM) \
608	606.9-608.2	N/A
609	608.5-609.3	Unc. (HI) \
616	615.4-616.2	N/A
627	626.9-627.7	Unc. (MDM) \ *Achromobacter* (MDM) \ Acetobacteraceae (MDM) \
628	627.7-628.5	*Achromobacter* (MDM) \
630	629.5-630.8	*Gammaproteobacteria* (MDM) \
632	631.6-632.4	*Dolosigranulum* (LOW) \
635	634.6-635.4	N/A
643	642.7-643.5	*Legionellaceae* (MDM) \
648	646.8-648.3	*Rhodococcus* (HI) \ *Bacteroidetes* (LOW) \ Clostridiales (LOW) \
651	650.3-651.1	N/A
651	650.8-652.1	Unc. (HI) \ *Carnobacteriaceae* (LOW) \
656	655.4-657	Bacteria (MDM) \ *Desulfosporosinus* (LOW) \
658	657.7-658.5	N/A
659	658.7-659.5	*Desulfosporosinus* (LOW) \
666	665.5-667.4	Bacteria (HI) \

(continued)

Table A.1 (Continued)

Allele	Bin Range	Matching Identity*
676	674.6-677.3	N/A
681	680.1-680.9	N/A
692	691.3-693.8	*Mycobacterium* (MDM) \
699	697.9-699.5	*Comamonadaceae* (HI) \ Bacteria (MDM) \ *Mycobacterium* (MDM) \
705	704.4-705.8	*Mycobacterium* (HI) \ Bacteria (HI) \ Unc. (MDM) \
706	706-706.8	Unc. (HI) \ *Mycobacterium* (MDM) \
707	707-707.8	N/A
711	709.9-711.1	*Azonexus* (MDM) \ *Flavobacterium* (MDM) \
712	710.9-712.7	*Mycobacterium* (HI) \ *Azonexus* (MDM) \ Unc. (MDM) \ *Flavobacterium* (MDM) \
714	713.4-714.8	*Mycobacterium* (HI) \ Unc. (LOW) \
716	715.2-717.3	*Mycobacterium* (MDM) \ *Bacteroidetes* (MDM) \ Unc. (MDM) \ *Porphyrobacter* (LOW) \
719	718-719.4	*Shigella* (MDM) \ *Sphingobacteriales* (MDM) \ *Mycobacterium* (MDM) \ Unc. (LOW) \
720	719.6-720.4	*Shigella* (MDM) \ *Mycobacterium* (MDM) \ Unc. (MDM) \
721	720.6-721.55	*Mycobacterium* (HI) \ *Gammaproteobacteria* (HI) \ Unc. (HI) \ Enterobacteriaceae (MDM) \
723	721.6-724	Unc. (HI) \ *Mycobacterium* (HI) \ *Gammaproteobacteria* (HI) \ *Enterobacteriaceae* (MDM) \
728	726.7-729.1	Unc. (HI) \ *Enterobacteriaceae* (MDM) \ *Gammaproteobacteria* (MDM) \ *Shigella* (MDM) \ *Mycobacterium* (MDM) \
732	731.2-733.1	*Silanimonas* (MDM) \ Bacteria (MDM) \
736	735.4-736.2	*Sphingobacteriales* (HI) \
740	739.4-740.2	Unc. (HI) \ *Gordonia* (HI) \
748	747.2-748	*Massilia* (MDM) \
749	748.5-750.2	Unc. (MDM) \ *Flavobacterium* (MDM) \ *Massilia* (MDM) \
756	754.6-756.6	*Rhodocyclaceae* (MDM) \ *Pseudomonas* (MDM) \ Unc. (LOW) \
759	757.9-759.4	*Mycobacterium* (HI) \ Unc. (MDM) \ *Nitrosomonas* (MDM) \
761	759.7-761.6	Unc. (MDM) \ *Nitrosomonas* (MDM) \
763	762.1-763.1	N/A
768	767.4-768.2	*Flavobacteriaceae* (MDM) \ *Bacteroidetes* (MDM) \
773	772.3-773.6	*Stenotrophomonas* (HI) \ *Acinetobacter* (MDM) \
775	774.2-776.3	Bacteria (HI) \ *Pseudomonas* (MDM) \ *Acinetobacter* (MDM) \ *Sphingobacterium* (LOW) \
780	778.5-780.4	*Rhodocyclaceae* (MDM) \ *Acidovorax* (MDM) \ *Pseudoxanthomonas* (MDM) \ *Sphingobacterium* (LOW) \

(continued)

Table A.1 (Continued)

Allele	Bin Range	Matching Identity*
782	782-782.8	*Stenotrophomonas* (MDM) \
784	782.9-785.4	*Stenotrophomonas* (HI) \ Unc. (MDM) \ *Sphingobacteriales* (MDM) \ *Flavobacterium* (MDM) \ Sphingomonadaceae (MDM) \ *Porphyrobacter* (LOW) \ *Sterolibacterium* (LOW) \
786	785.6-787.2	*Flavobacterium* (HI) \ *Sphingobacteriales* (MDM) \ *Pseudomonas* (MDM) \ *Gammaproteobacteria* (MDM) \
788	787.2-788.3	*Gammaproteobacteria* (MDM) \ *Pseudomonas* (MDM) \
788	788.1-788.9	Pseudomonas (MDM) \ *Gammaproteobacteria* (MDM) \ *Nitrospira* (MDM) \
790	789.3-792	*Nitrospira* (HI) \ *Flavobacterium* (HI) \ *Pseudomonas* (MDM) \ *Gammaproteobacteria* (MDM) \ *Azonexus* (MDM) \ *Sphingobacteriales* (MDM) \ *Rubrivivax* (LOW) \ Unc. (LOW) \ *Bacteroidetes* (LOW) \
794	794-794.8	*Rubrivivax* (LOW) \
796	795.9-796.7	*Sphingobacteriales* (HI) \ Unc. (MDM) \ *Nitrosomonas* (MDM) \ *Pseudomonas* (LOW) \
798	796-798.7	*Sphingobacteriales* (HI) \ *Pseudomonas* (HI) \ Unc. (MDM) \ *Nitrosomonas* (MDM) \ *Sphingobacterium* (MDM) \
799	798.8-799.6	*Nitrosomonas* (MDM) \ *Pseudomonas* (MDM) \
801	800.3-801.1	*Sphingobacterium* (MDM) \
803	802.6-803.4	*Azonexus* (HI) \
805	804.3-805.1	N/A
807	805.4-807.3	Unc. (HI) \ *Alphaproteobacteria (*HI) \ *Flavobacterium* (LOW) \
808	807.7-809.5	*Pseudomonas* (MDM) \ *Dysgonomonas* (MDM) \ *Bacteroidetes* (LOW) \ *Mycobacterium* (LOW) \
811	810.1-813.7	*Arthrobacter* (HI) \ *Dysgonomonas* (MDM) \ *Bacteroidetes* (MDM) \ Sphingobacteriales (MDM) \ *Pseudomonas* (MDM) \ *Rheinheimera* (MDM) \ Unc. (MDM) \ *Flavobacterium* (LOW) \ *Janthinobacterium* (LOW) \
813	812.7-814.1	*Bacteroidetes* (MDM) \ *Rheinheimera* (MDM) \
819	817.5-819.6	Xanthomonadaceae (HI) \ *Stenotrophomonas* (HI) \ Sphingobacteriales (HI) \ *Chryseobacterium* (HI) \ Unc. (MDM) \ *Legionella* (MDM) \ Flexibacteraceae (MDM) \
821	819.6-821.7	*Sphingobacteriales* (HI) \ *Shigella* (HI) \ Unc. (MDM) \ *Legionella* (MDM) \
827	825.2-827.3	Unc. (HI) \ *Alphaproteobacteria* (HI) \ *Gammaproteobacteria* (LOW) \
829	828.1-830.4	*Nitrospira* (HI) \ *Nitrosomonas* (HI) \ *Alphaproteobacteria* (HI) \ *Emticicia* (MDM) \ *Stenotrophomonas* (LOW) \
832	830.9-833.4	*Sphingomonadaceae* (LOW) \ *Stenotrophomonas* (LOW) \

(continued)

Table A.1 (Continued)

Allele	Bin Range	Matching Identity*
843	842.5-843.7	*Sphingomonadaceae* (MDM) \ *Sphingomonas* (LOW) \
847	845.6-847.4	*Gammaproteobacteria* (HI) \ *Arthrobacter* (MDM) \ Unc. (MDM) \ *Sphingomonas* (LOW) \
848	845.9-849.2	*Arthrobacter* (HI) \ *Gammaproteobacteria* (HI) \ Unc. (MDM) \
852	850.7-853.2	*Proteobacteria* (MDM) \ Gammaproteobacteria (MDM) \ *Betaproteobacteria* (MDM) \ *Nitrosomonas* (LOW) \
856	855.1-856.6	Unc. (HI) \ *Sphingobacteriales* (MDM) \ *Betaproteobacteria* (MDM) \
861	860.1-862.6	Unc. (HI) \ *Betaproteobacteria* (MDM) \
863	862.3-865.5	*Gammaproteobacteria* (MDM) \ *Alphaproteobacteria* (MDM) \ *Burkholderiales* (MDM) \ *Betaproteobacteria* (MDM) \
867	865.7-867.5	*Flavobacteriaceae* (MDM) \ *Gammaproteobacteria* (MDM) \ *Alphaproteobacteria* (MDM) \ *Flavobacterium* (LOW) \
869	867.2-870.6	*Variovorax* (HI) \ *Chryseobacterium* (HI) \ *Flavobacteriaceae* (MDM) \ Unc. (MDM) \ Bacteria (MDM) \
872	871.2-873.4	*Variovorax* (HI) \ *Chryseobacterium* (HI) \ Unc. (MDM) \ *Brevundimonas* (MDM) \ *Sphingomonas* (MDM) \
876	875-877.2	*Pseudomonas* (HI) \ *Methylophilus* (HI) \ Unc. (MDM) \ *Arcicella* (MDM) \
878	877.3-879.4	N/A
881	879.7-883.5	*Pseudomonas* (HI) \ Alcaligenaceae (LOW) \
888	886-889.4	Unc. (HI) \ *Sphingobacteriales* (HI) \ *Curvibacter* (HI) \ *Oxalobacteraceae* (MDM) \ *Comamonadaceae* (MDM) \ *Pseudomonas* (LOW) \ *Alcanivorax* (LOW) \
892	890.6-892.1	*Xanthomonadaceae* (MDM) \ *Sphingomonadaceae* (MDM) \ *Gammaproteobacteria* (LOW) \
893	892.2-894.8	*Sphingopyxis* (HI) \ *Brevundimonas* (MDM) \ Unc. (MDM) \ *Sphingomonadaceae* (MDM) \
899	898-900	*Alphaproteobacteria* (HI) \ *Methylomonas* (MDM) \ Unc. (MDM) \ *Methylococcaceae* (MDM) \ *Novosphingobium* (MDM) \ *Sphingomonas* (LOW) \
901	900.7-901.7	*Sphingomonadaceae* (HI) \ *Sphingomonas* (MDM) \ Unc. (MDM) \
902	900.6-904.6	*Sphingomonadaceae* (HI) \ *Methylomonas* (MDM) \ Unc. (MDM) \ *Sphingomonas* (MDM) \ *Methylococcaceae* (MDM) \ *Flavobacterium* (MDM) \
912	911.2-913.1	*Acinetobacter* (MDM) \ *Methylobacterium* (MDM) \ *Sphingomonadaceae* (MDM) \
915	913.9-916.1	*Methylobacterium* (MDM) \
917	916.1-917.9	*Sphingomonadaceae* (MDM) \ *Sphingomonas* (MDM) \

(continued)

Table A.1 (Continued)

Allele	Bin Range	Matching Identity*
920	919.2-921.2	*Sphingomonadaceae* (HI) \ *Sphingomonas* (HI) \
922	921.3-924.1	Unc. (HI) \ *Sphingopyxis* (HI) \ *Sphingomonadaceae* (HI) \ *Sphingomonas* (MDM) \ *Methylophilus* (MDM) \ *Chryseobacterium* (MDM) \ *Alphaproteobacteria (*MDM) \
925	924.2-925.3	*Acinetobacter* (MDM) \ *Alphaproteobacteria* (MDM) \ Unc. (MDM) \
926	925.7-926.5	*Acinetobacter* (MDM) \
928	928-928.8	Unc. (HI) \ *Nitrosomonas* (HI) \ *Acinetobacter* (MDM) \
930	927.4-931.8	Unc. (HI) \ *Nitrosomonas* (HI) \ *Rhodoferax* (HI) \ *Phenylobacterium* (HI) \ *Sphingopyxis* (HI) \ *Acinetobacter* (MDM) \
931	930.1-932.9	*Rhodoferax* (HI) \ Gp3 (MDM) \
939	937.7-939.6	*Acidovorax* (HI) \ *Alphaproteobacteria* (MDM) \ *Sphingomonadaceae* (MDM) \
941	938.1-942.6	*Acidovorax* (HI) \ *Acinetobacter* (MDM) \ *Alphaproteobacteria* (MDM) \ *Proteobacteria* (MDM) \ *Sphingomonadaceae* (MDM) \
949	948.2-949	N/A
950	949.7-950.5	N/A
953	951.5-955.8	*Caulobacter* (HI) \ *Flexibacteraceae* (MDM) \ *Sphingomonadaceae* (MDM) \ *Sphingomonas* (MDM) \
960	959.5-960.9	Unc. (HI) \ *Caulobacter* (HI) \ *Arcicella* (MDM) \ *Cohnella* (MDM) \
963	961.9-965.3	*Sphingopyxis* (HI) \ Unc. (HI) \ *Gammaproteobacteria* (MDM) \ *Arcicella* (MDM) \ *Cohnella* (MDM) \ *Sphingomonadaceae* (LOW) \
969	967.3-971.1	*Caulobacter* (HI) \ *Acinetobacter* (HI) \
974	971.9-976	*Sphingomonadaceae* (HI) \ *Phenylobacterium* (HI) \ Hyphomicrobiaceae (HI) \ *Gammaproteobacteria* (HI) \ *Bacteroidetes* (HI) \ *Alphaproteobacteria* (HI) \ *Caulobacter* (HI) \ *Acinetobacter* (HI) \ *Brucellaceae* (MDM) \
978	976.9-978.3	*Caulobacter* (HI) \ *Pseudochrobactrum* (HI) \ *Alcaligenaceae* (LOW) \ *Oxalobacteraceae* (LOW) \
981	978.7-981	Unc. (HI) \ *Sphingopyxis* (HI) \ *Caulobacter* (HI) \ *Gammaproteobacteria* (LOW) \ *Oxalobacteraceae* (LOW) \
982	979-984.3	Unc. (HI) \ *Sphingopyxis* (HI) \ *Pseudochrobactrum* (HI) \ *Chryseobacterium* (MDM) \ *Gammaproteobacteria* (LOW) \ *Oxalobacteraceae* (LOW) \
983	982.5-985	Unc. (HI) \ *Pseudochrobactrum* (HI) \ *Chryseobacterium* (MDM) \

(continued)

Table A.1 (Continued)

Allele	Bin Range	Matching Identity*
987	985.3-990	*Sphingomonadaceae* (HI) \ *Pedobacter* (HI) \ *Hyphomicrobiaceae* (MDM) \
994	992.4-994.7	Unc. (HI) \ *Sphingomonadaceae* (HI) \ *Gammaproteobacteria* (HI) \ *Bdellovibrio* (HI) \ *Alphaproteobacteria* (HI) \ *Porphyrobacter* (MDM) \ *Hyphomicrobiaceae* (MDM) \
997	994.8-999.3	*Sphingopyxis* (HI) \ *Bosea* (HI) \ *Alphaproteobacteria* (HI) \ *Hyphomicrobiaceae* (MDM) \
1002	1000.7-1004.4	*Caulobacter* (HI) \ *Hyphomicrobium* (HI) \ *Acinetobacter* (MDM) \ *Limnobacter* (MDM) \ *Desulfosporosinus* (MDM) \ Unc. (MDM) \ *Blastomonas* (MDM) \ *Flavobacteriaceae* (LOW) \
1003	1002.5-1004.4	*Caulobacter* (HI) \ *Hyphomicrobium* (HI) \ *Limnobacter* (MDM) \ Unc. (MDM) \ *Blastomonas* (MDM) \
1007	1005.5-1007.5	Unc. (HI) \ *Sphingopyxis* (HI) \
1012	1010.6-1012.1	Unc. (HI) \ *Sphingomonadaceae* (HI) \ *Phenylobacterium* (HI) \ *Hyphomicrobiaceae* (HI) \ *Alphaproteobacteria* (HI) \ *Rhizobiales* (MDM) \
1013	1010.7-1014.2	Unc. (HI) \ *Sphingomonadaceae* (HI) \ *Phenylobacterium* (HI) \ *Hyphomicrobiaceae* (HI) \ *Alphaproteobacteria* (HI) \ *Rhizobiales* (MDM) \
1018	1016.4-1018.8	Unc. (HI) \ *Alphaproteobacteria* (HI) \ *Methylobacterium* (HI) \ *Porphyrobacter* (LOW) \
1021	1020.1-1021.6	Unc. (HI) \ *Methylobacterium* (HI) \ Bacteria (HI) \ *Hyphomicrobium* (HI) \ *Proteobacteria* (MDM) \ *Limnobacter* (MDM) \ *Gammaproteobacteria* (MDM) \
1024	1022.1-1025.7	*Porphyrobacter* (HI) \ *Hyphomicrobium* (HI) \ *Hyphomicrobiaceae* (HI) \ *Methylobacterium* (HI) \ *Limnobacter* (MDM) \ Unc. (LOW) \
1026	1026-1026.8	*Novosphingobium* (HI) \ Hyphomicrobiaceae (HI) \ Gp3 (HI) \ *Alphaproteobacteria* (HI) \
1033	1032.3-1033.8	Unc. (HI) \ *Novosphingobium* (HI) \ *Caulobacter* (HI) \ *Sphingomonas* (MDM) \
1034	1033.9-1035.4	*Novosphingobium* (HI) \ Alphaproteobacteria (HI) \ *Bradyrhizobium* (MDM) \ Unc. (LOW) \
1037	1035.5-1037.5	*Bradyrhizobium* (MDM) \
1038	1037.7-1039.1	*Rhizobiales* (HI) \ *Limnobacter* (HI) \ *Bradyrhizobiaceae* (HI) \
1041	1039.2-1041.7	*Bradyrhizobiaceae* (MDM) \ *Alphaproteobacteria* (MDM) \
1042	1041.7-1043	*Limnobacter* (HI) \
1044	1043.4-1044.2	*Limnobacter* (HI) \
1046	1044.7-1049	*Sphingomonas* (HI) \ *Methylophilus* (HI) \ *Afipia* (HI) \ *Alphaproteobacteria* (MDM) \

(continued)

Table A.1 (Continued)

Allele	Bin Range	Matching Identity*
1054	1050-1057	*Roseomonas* (HI) \ *Alphaproteobacteria* (HI) \ *Sphingomonas* (MDM) \ *Bradyrhizobiaceae* (MDM) \
1058	1057.5-1058.3	N/A
1064	1058.4-1065.9	N/A

*16S identities of matching clones based on a 95% confidence, according to RDP Classifier (Cole et al. 2005). Script in parentheses indicate the level of confidence in the clone assignments to the allele: HI, high; MDM, medium; LOW, low. Unc., unclassified; N/A, not available.

REFERENCES

Abraham, W. R., C. Strompl, H. Meyer, S. Lindholst, E. R. Moore, R. Christ, M. Vancanneyt, B. J. Tindall, A. Bennasar, J. Smit and M. Tesar. 1999. Phylogeny and polyphasic taxonomy of Caulobacter species. Proposal of Maricaulis gen. nov. with Maricaulis maris (Poindexter) comb. nov. as the type species, and emended description of the genera Brevundimonas and Caulobacter. *International Journal of Systematic and Evolutionary Microbiology,* 49 Pt 3:1053-1073.

Amann, R. and W. Ludwig. 2000. Ribosomal RNA-targeted nucleic acid probes for studies in microbial ecology. *FEMS Microbiology Reviews,* 24:555-565.

Amann, R., W. Ludwig and K.-H. Schleifer. 1995. Phylogenetic Identification and In Situ Detection of Individual Microbial Cells without Cultivation. *Microbiological Reviews,* 59(1):143-169.

AWWA. 2000. Committee report: Disinfection at large and medium-size systems. *Journal American Water Works Association,* 92:32-43.

AWWA. 2006. *Fundamentals and Control of Nitrification in Chloraminated Drinking Water Distribution Systems*. Denver, CO. American Water Works Association.

Benson, D. A., I. Karsch-Mizrachi, D. J. Lipman, J. Ostell and D. L. Wheeler. 2006. GenBank. *Nucleic Acids Research,* 34(Database issue):D16-20.

Berekaa, M. M. and A. Steinbuchel. 2000. Microbial degradation of the multiply branched alkane 2,6,10,15,19,23-hexamethyltetracosane (squalane) by Mycobacterium fortuitum and Mycobacterium ratisbonense. *Applied and Environmental Microbiology,* 66(10):4462-4467.

Berry, D., C. Xi and L. Raskin. 2006. Microbial ecology of drinking water distribution systems. *Current Opinion in Biotechnology,* 17(3):297-302.

Brodtmann, N. V. and P. J. Russo. 1979. Use of Chloramine for Reduction of Trihalomethanes and Disinfection of Drinking-Water. *Journal American Water Works Association,* 71(1):40-42.

Brown, M. V., M. S. Schwalbach, I. Hewson and J. A. Fuhrman. 2005. Coupling 16S-ITS rDNA clone libraries and automated ribosomal intergenic spacer analysis to show marine microbial diversity: development and application to a time series. *Environmental Microbiology,* 7(9):1466-1479.

Bruce, K. D. 1997. Analysis of mer gene subclasses within bacterial communities in soils and sediments resolved by Fluorescent-PCR-restriction fragment length polymorphism profiling. *Applied and Environmental Microbiology,* 63(12):4914-4919.

Burland, V., G. Plunkett, D. L. Daniels and F. R. Blattner. 1993. DNA-Sequence and Analysis of 136 Kilobases of the Escherichia-Coli Genome - Organizational Symmetry around the Origin of Replication. *Genomics,* 16(3):551-561.

Cardinale, M., L. Brusetti, P. Quatrini, S. Borin, A. M. Puglia, A. Rizzi, E. Zanardini, C. Sorlini, C. Corselli and D. Daffonchio. 2004. Comparison of different primer sets for use in automated ribosomal intergenic spacer analysis of complex bacterial communities. *Applied and Environmental Microbiology,* 70(10):6147-6156.

Carrero-Colon, M., C. H. Nakatsu and A. Konopka. 2006. Effect of nutrient periodicity on microbial community dynamics. *Applied and Environmental Microbiology,* 72(5):3175-3183.

Carson, L. A., N. J. Petersen, M. S. Favero and S. M. Aguero. 1978. Growth characteristics of atypical mycobacteria in water and their comparative resistance to disinfectants. *Applied and Environmental Microbiology,* 36(6):839-846.

Chou, H. H. and M. H. Holmes. 2001. DNA sequence quality trimming and vector removal. *Bioinformatics,* 17:1093-1104.

Cilia, V., B. Lafay and R. Christen. 1996. Sequence heterogeneities among 16S ribosomal RNA sequences, and their effect on phylogenetic analyses at the species level. *Molecular Biology and Evolution,* 13(3):451-461.

Cole, J. R., B. Chai, R. J. Farris, Q. Wang, A. S. Kulam-Syed-Mohideen, D. M. McGarrell, A. M. Bandela, E. Cardenas, G. M. Garrity and J. M. Tiedje. 2007. The ribosomal database project (RDP-II): introducing myRDP space and quality controlled public data. *Nucleic Acids Research,* 35:D169-D172.

Cole, J. R., B. Chai, R. J. Farris, Q. Wang, S. A. Kulam, D. M. McGarrell, G. M. Garrity and J. M. Tiedje. 2005. The Ribosomal Database Project (RDP-II): sequences and tools for high-throughput rRNA analysis. *Nucleic Acids Research,* 33(Database issue):D294-296.

Connon, S. A., A. Tovanabootr, M. Dolan, K. Vergin, S. J. Giovannoni and L. Semprini. 2005. Bacterial community composition determined by culture-independent and -dependent methods during propane-stimulated bioremediation in trichloroethene-contaminated groundwater. *Environmental Microbiology,* 7(2):165-178.

Cunliffe, D. A. 1991. Bacterial Nitrification in Chloraminated Water-Supplies. *Applied and Environmental Microbiology,* 57(11):3399-3402.

Dailloux, M., M. Albert, C. Laurain, S. Andolfatto, A. Lozniewski, P. Hartemann and L. Mathieu. 2003. Mycobacterium xenopi and drinking water biofilms. *Applied and Environmental Microbiology,* 69(11):6946-6948.

de Hoon, M. J. L., S. Imoto, J. Nolan and S. Miyano. 2004. Open source clustering software. *Bioinformatics,* 20(9):1453-1454.

Eisen, M. B., P. T. Spellman, P. O. Brown and D. Botstein. 1998. Cluster analysis and display of genome-wide expression patterns. *Proceedings of the National Academy of Sciences of the United States of America,* 95(25):14863-14868.

Ewing, B. and P. Green. 1998. Base-calling of automated sequencer traces using phred. II. Error probabilities. *Genome Research,* 8(3):186-194.

Ewing, B., L. Hillier, M. C. Wendl and P. Green. 1998. Base-calling of automated sequencer traces using phred. I. Accuracy assessment. *Genome Research,* 8(3):175-185.

Falkinham, J. O., 3rd, C. D. Norton and M. W. LeChevallier. 2001. Factors influencing numbers of Mycobacterium avium, Mycobacterium intracellulare, and other Mycobacteria in drinking water distribution systems. *Applied and Environmental Microbiology,* 67(3):1225-1231.

Feurer, C., D. Clermont, F. Bimet, A. Candrea, M. Jackson, P. Glaser, C. Bizet and C. Dauga. 2004. Taxonomic characterization of nine strains isolated from clinical and environmental specimens, and proposal of Corynebacterium tuberculostearicum sp nov. *International Journal of Systematic and Evolutionary Microbiology,* 54:1055-1061.

Fisher, M. M. and E. W. Triplett. 1999. Automated approach for ribosomal intergenic spacer analysis of microbial diversity and its application to freshwater bacterial communities. *Applied and Environmental Microbiology,* 65(10):4630-4636.

Fleming, K. K., G. W. Harrington and D. R. Noguera. 2005. Nitrification potential curves: a new strategy for nitrification prevention. *Journal American Water Works Association,* 97(8):90-99.

Flemming, L., D. Rawlings and H. Chenia. 2007. Phenotypic and molecular characterisation of fish-borne Flavobacterium johnsoniae-like isolates from aquaculture systems in South Africa. *Research in Microbiology,* 158(1):18-30.

Forney, L. J., X. Zhou and C. J. Brown. 2004. Molecular microbial ecology: land of the one-eyed king. *Current Opinion in Microbiology,* 7(3):210-220.

Fritz, I., C. Strompl, D. I. Nikitin, A. M. Lysenko and W. R. Abraham. 2005. Brevundimonas mediterranea sp nov., a non-stalked species from the Mediterranean Sea. *International Journal of Systematic and Evolutionary Microbiology,* 55:479-486.

Gagnon, G. A., H. M. Murphy, J. L. Rand, S. J. Payne, S. Springthorpe, F. Matias, R. Nokbeh and S. A. Sattar. 2007. *Coliforms in Distribution Systems: Integrated Disinfection and Antimicrobial Resistance*. AWWA Research Foundation.

Gallego, V., M. T. Garcia and A. Ventosa. 2005. Methylobacterium hispanicum sp nov and Methylobacterium aquaticum sp nov., isolated from drinking water. *International Journal of Systematic and Evolutionary Microbiology,* 55:281-287.

Gallego, V., M. T. Garcia and A. Ventosa. 2006. Methylobacterium adhaesivum sp nov, a methylotrophic bacterium isolated from drinking water. *International Journal of Systematic and Evolutionary Microbiology,* 56:339-342.

Garcia-Martinez, J., I. Bescos, J. J. Rodriguez-Sala and F. Rodriguez-Valera. 2001. RISSC: a novel database for ribosomal 16S-23S RNA genes spacer regions. *Nucleic Acids Research,* 29(1):178-180.

Giavannoni, S. J. 1991. The polymerase chain reaction.In *Nucleic acid techniques in bacterial systematics*. Edited by E. Stackebrandt and M. Goodfellow. New York, N.Y., John Wiley & Sons: 177-201.

Gill, S. R., D. E. Fouts, G. L. Archer, E. F. Mongodin, R. T. DeBoy, J. Ravel, I. T. Paulsen, J. F. Kolonay, L. Brinkac, M. Beanan, R. J. Dodson, S. C. Daugherty, R. Madupu, S. V. Angiuoli, A. S. Durkin, D. H. Haft, J. Vamathevan, H. Khouri, T. Utterback, C. Lee, G. Dimitrov, L. X. Jiang, H. Y. Qin, J. Weidman, K. Tran, K. Kang, I. R. Hance, K. E. Nelson and C. M. Fraser. 2005. Insights on evolution of virulence and resistance from the complete genome analysis of an early methicillin-resistant Staphylococcus aureus strain and a biofilm-producing methicillin-resistant Staphylococcus epidermidis strain. *Journal of Bacteriology,* 187(7):2426-2438.

Glaze, W. H. 1990. Chemical Oxidation.In *Water Quality and Treatment, A Handbook of Community Water Supplies*. New York, McGraw-Hill, Inc.

Hall, T. A. 1999. BioEdit: a user-friendly biological sequence alignment editor and analysis program for Windows 95/98/NT. *Nucleic Acids Symposium Series,* 41:95-98.

Harrington, G. W., D. R. Noguera, A. I. Kandou, P. S. Oldenburg, J. M. Regan and D. J. VanHoven. 2003. Ammonia release from chloramine decay: Implications for the prevention and control of nitrification in distribution systems. *AWWA Research Foundation,* ISBN 1-58321-58283-58323.

Harrington, G. W., D. R. Noguera, A. I. Kandou and D. J. Vanhoven. 2002. Pilot-scale evaluation of nitrification control strategies. *Journal American Water Works Association,* 94(11):78-89.

Hiraishi, A. and M. Kaneko. 1994. Use of polymerase chain reaction-amplified 16S rRNA gene sequences to identify pink-pigmented bacteria found in a potable water treatment system. *Bulletin of the Japanese Society for Microbial Ecology,* 9:55-65.

Hoefel, D., P. T. Monis, W. L. Grooby, S. Andrews and C. P. Saint. 2005. Profiling bacterial survival through a water treatment process and subsequent distribution system. *Journal of Applied Microbiology,* 99(1):175-186.

Hoffmann, A., T. Thimm, M. Droge, E. R. B. Moore, J. C. Munch and C. C. Tebbe. 1998. Intergeneric transfer of conjugative and mobilizable plasmids harbored by Escherichia coli in the gut of the soil microarthropod Folsomia candida (Collembola). *Applied and Environmental Microbiology,* 64(7):2652-2659.

Huber, T., G. Faulkner and P. Hugenholtz. 2004. Bellerophon: a program to detect chimeric sequences in multiple sequence alignments. *Bioinformatics,* 20(14):2317-2319.

Inoue, K., H. Habe, H. Yamane, T. Omori and H. Nojiri. 2005. Diversity of carbazole-degrading bacteria having the car gene cluster: Isolation of a novel gram-positive carbazole-degrading bacterium. *FEMS Microbiology Letters,* 245(1):145-153.

Ivanova, E. P., N. M. Gorshkova, T. Sawabe, K. Hayashi, N. I. Kalinovskaya, A. M. Lysenko, N. V. Zhukova, D. V. Nicolau, T. A. Kuznetsova, V. V. Mikhailov and R. Christen. 2002. Pseudomonas extremorientalis sp nov., isolated from a drinking water reservoir. *International Journal of Systematic and Evolutionary Microbiology,* 52:2113-2120.

Jones, S. E., A. L. Shade, K. D. McMahon and A. D. Kent. 2007. Comparison of primer sets for use in automated ribosomal intergenic spacer analysis of aquatic bacterial communities: an ecological perspective. *Applied and Environmental Microbiology,* 73(2):659-662.

Kaiser, O., A. Puhler and W. Selbitschka. 2001. Phylogenetic analysis of microbial diversity in the rhizoplane of oilseed rape (Brassica napus cv. Westar) employing cultivation-dependent and cultivation-independent approaches. *Microbial Ecology,* 42(2):136-149.

Kalmbach, S., W. Manz, J. Wecke and U. Szewzyk. 1999. Aquabacterium gen. nov., with description of Aquabacterium citratiphilum sp. nov., Aquabacterium parvum sp. nov. and Aquabacterium commune sp. nov., three in situ dominant bacterial species from the Berlin drinking water system. *International Journal of Systematic Bacteriology,* 49:769-777.

Kent, A. D., D. J. Smith, B. J. Benson and E. W. Triplett. 2003. Web-based phylogenetic assignment tool for analysis of terminal restriction fragment length polymorphism profiles of microbial communities. *Applied and Environmental Microbiology,* 69(11):6768-6776.

Kim, Y. H., K. H. Engesser and S. J. Kim. 2007. Physiological, numerical and molecular characterization of alkyl ether-utilizing rhodococci. *Environmental Microbiology,* 9(6):1497-1510.

Klappenbach, J. A., P. R. Saxman, J. R. Cole and T. M. Schmidt. 2001. rrndb: the Ribosomal RNA Operon Copy Number Database. *Nucleic Acids Research,* 29(1):181-184.

Kruskal, J. B. 1964. Multidimensional scaling by optimizing goodness of fit to a nonmetric hypothesis. *Psychometrika,* 29(1-27.

La Duc, M. T., W. Nicholson, R. Kern and K. Venkateswaran. 2003. Microbial characterization of the Mars Odyssey spacecraft and its encapsulation facility. *Environmental Microbiology,* 5(10):977-985.

Lane, D. J. 1991. 16S/23S rRNA sequencing.In *Nucleic acid techniques in bacterial systematics.* Edited by E. Stackebrandt and M. Goodfellow. New York, N.Y., John Wiley and Sons: 115-148.

Lane, D. J., B. Pace, G. J. Olsen, D. A. Stahl, M. L. Sogin and N. R. Pace. 1985. Rapid-Determination of 16S Ribosomal-Rna Sequences for Phylogenetic Analyses. *Proceedings of the National Academy of Sciences of the United States of America,* 82(20):6955-6959.

Larsen, P., J. L. Nielsen, M. S. Dueholm, R. Wetzel, D. Otzen and P. H. Nielsen. 2007. Amyloid adhesins are abundant in natural biofilms. *Environmental Microbiology,* 9(12):3077-3090.

Legendre, P. and L. Legendre. 1998. *Numerical ecology, 2nd English ed.* Amsterdam, The Netherlands. Elsevier Science BV.

Leser, T. D., J. Z. Amenuvor, T. K. Jensen, R. H. Lindecrona, M. Boye and K. Moller. 2002. Culture-independent analysis of gut bacteria: the pig gastrointestinal tract microbiota revisited. *Applied and Environmental Microbiology,* 68(2):673-690.

Lieu, N. I., R. L. Wolfe and E. G. Means. 1993. Optimizing Chloramine Disinfection for the Control of Nitrification. *Journal American Water Works Association,* 85(2):84-90.

Lin, L. H., P. L. Wang, D. Rumble, J. Lippmann-Pipke, E. Boice, L. M. Pratt, B. S. Lollar, E. L. Brodie, T. C. Hazen, G. L. Andersen, T. Z. DeSantis, D. P. Moser, D. Kershaw and T. C. Onstott. 2006. Long-term sustainability of a high-energy, low-diversity crustal biome. *Science,* 314(5798):479-482.

Liu, W. T., T. L. Marsh, H. Cheng and L. J. Forney. 1997. Characterization of microbial diversity by determining terminal restriction fragment length polymorphisms of genes encoding 16S rRNA. *Applied and Environmental Microbiology,* 63(11):4516-4522.

Mergaert, J., A. Boley, M. C. Cnockaert, W. R. Muller and J. Swings. 2001. Identity and potential functions of heterotrophic bacterial isolates from a continuous-upflow fixed-bed reactor for denitrification of drinking water with bacterial polyester as source of carbon and electron donor. *Systematic and Applied Microbiology,* 24(2):303-310.

Moissl, C., S. Osman, M. T. La Duc, A. Dekas, E. Brodie, T. DeSantis and K. Venkateswaran. 2007. Molecular bacterial community analysis of clean rooms where spacecraft are assembled. *FEMS Microbiology Ecology,* 61(3):509-521.

Muyzer, G., E. C. Dewaal and A. G. Uitterlinden. 1993. Profiling of Complex Microbial-Populations by Denaturing Gradient Gel-Electrophoresis Analysis of Polymerase Chain Reaction-Amplified Genes-Coding for 16s Ribosomal-Rna. *Applied and Environmental Microbiology,* 59(3):695-700.

Newton, R. J., A. D. Kent, E. W. Triplett and K. D. McMahon. 2006. Microbial community dynamics in a humic lake: differential persistence of common freshwater phylotypes. *Environmental Microbiology,* 8(6):956-970.

Norton, C. D. and M. W. LeChevallier. 1997. Chloramination: Its effect on distribution system water quality. *Journal American Water Works Association,* 89(7):66-77.

Odell, L. H., G. J. Kirmeyer, A. Wilczak, J. G. Jacangelo, J. P. Marcinko and R. L. Wolfe. 1996. Controlling nitrification in chloraminated systems. *Journal American Water Works Association,* 88(7):86-98.

Oldenburg, P. S., J. M. Regan, G. W. Harrington and D. R. Noguera. 2002. Kinetics of Nitrosomonas europaea inactivation by chloramine. *Journal American Water Works Association,* 94(10):100-110.

Olsen, G. J. and C. R. Woese. 1993. Ribosomal RNA: a key to phylogeny. *FASEB Journal,* 7(1):113-123.

Pernthaler, J., T. Posch, K. Simek, J. Vrba, A. Pernthaler, F. O. Glockner, U. Nubel, R. Psenner and R. Amann. 2001. Predator-specific enrichment of actinobacteria from a cosmopolitan freshwater clade in mixed continuous culture. *Applied and Environmental Microbiology,* 67(5):2145-2155.

Pfaller, S. L., T. W. Aronson, A. E. Holtzman and T. C. Covert. 2007. Amplified fragment length polymorphism analysis of Mycobacterium avium complex isolates recovered from southern California. *Journal of Medical Microbiology,* 56(9):1152-1160.

Pintar, K. D. M. and R. M. Slawson. 2003. Effect of temperature and disinfection strategies on ammonia-oxidizing bacteria in a bench-scale drinking water distribution system. *Water Research,* 37(8):1805-1817.

Pryor, M., S. Springthorpe, S. Riffard, T. Brooks, Y. Huo, G. Davis and S. A. Satter. 2004. Investigation of opportunistic pathogens in municipal drinking water under different supply and treatment regimes. *Water Science and Technology,* 50(1):83-90.

Regan, J. M. (2001). Microbial Ecology of Nitrification in Chloraminated Drinking Water Distribution Systems. Civil and Environmental Engineering. Madison, WI, University of Wisconsin - Madison.

Regan, J. M., G. W. Harrington, H. Baribeau, R. De Leon and D. R. Noguera. 2003. Diversity of nitrifying bacteria in full-scale chloraminated distribution systems. *Water Research,* 37(1):197-205.

Regan, J. M., G. W. Harrington and D. R. Noguera. 2002. Ammonia- and nitrite-oxidizing bacterial communities in a pilot-scale chloraminated drinking water distribution system. *Applied and Environmental Microbiology,* 68(1):73-81.

Ronquist, F. and J. P. Huelsenbeck. 2003. MrBayes 3: Bayesian phylogenetic inference under mixed models. *Bioinformatics,* 19(12):1572-1574.

Saiki, R. K., D. H. Gelfand, S. Stoffel, S. J. Scharf, R. Higuchi, G. T. Horn, K. B. Mullis and H. A. Erlich. 1988. Primer-directed enzymatic amplification of DNA with a thermostable DNA polymerase. *Science,* 239(4839):487-491.

Schloss, P. D. and J. Handelsman. 2005. Introducing DOTUR, a computer program for defining operational taxonomic units and estimating species richness. *Applied and Environmental Microbiology,* 71(3):1501-1506.

Schulze, R., S. Spring, R. Amann, I. Huber, W. Ludwig, K. H. Schleifer and P. Kampfer. 1999. Genotypic diversity of Acidovorax strains isolated from activated sludge and description of Acidovorax defluvii sp nov. *Systematic and Applied Microbiology,* 22(2):205-214.

Scupham, A. J. 2007. Examination of the microbial ecology of the avian intestine in vivo using bromodeoxyuridine. *Environmental Microbiology,* 9(7):1801-1809.

Skadsen, J. 1993. Nitrification in a Distribution-System. *Journal American Water Works Association,* 85(7):95-103.

Smith, R. L., S. P. Buckwalter, D. A. Repert and D. N. Miller. 2005. Small-scale, hydrogen-oxidizing-denitrifying bioredctor for treatment of nitrate-pontaminated drinking water. *Water Research,* 39(10):2014-2023.

Taylor, R. H., J. O. Falkinham, 3rd, C. D. Norton and M. W. LeChevallier. 2000. Chlorine, chloramine, chlorine dioxide, and ozone susceptibility of Mycobacterium avium. *Applied and Environmental Microbiology,* 66(4):1702-1705.

Tokajian, S. T., F. A. Hashwa, I. C. Hancock and P. A. Zalloua. 2005. Phylogenetic assessment of heterotrophic bacteria from a water distribution system using 16S rDNA sequencing. *Canadian Journal of Microbiology,* 51(4):325-335.

Tsitko, I., R. Rahkila, O. Priha, T. Ali-Vehmas, Z. Terefework, H. Soini and M. S. Salkinoja-Salonen. 2006. Isolation and automated ribotyping of Mycobacterium lentiflavum from drinking water distribution system and clinical specimens. *FEMS Microbiology Letters,* 256(2):236-243.

USEPA. 1989. Disinfection; turbidity, Giardia lamblia, viruses, Legionella, and heterotrophic bacteria; final rule.

USEPA. 1989. Total coliforms (including fecal coliforms and E. coli). Final rule.

USEPA. 1991. SOCs and IOCs. Final rule.

USEPA. 1998. Disinfectants and disinfection by-products; final rule.

Vallaeys, T., E. Topp, G. Muyzer, V. Macheret, G. Laguerre, A. Rigaud and G. Soulas. 1997. Evaluation of denaturing gradient gel electrophoresis in the detection of 16S rDNA sequence variation in rhizobia and methanotrophs. *FEMS Microbiology Ecology,* 24(3):279-285.

Van Trappen, S., J. Mergaert, S. Van Eygen, P. Dawyndt, M. C. Cnockaert and J. Swings. 2002. Diversity of 746 heterotrophic bacteria isolated from microbial mats from ten Antarctic lakes. *Systematic and Applied Microbiology,* 25(4):603-610.

von Wintzingerode, F., U. B. Gobel and E. Stackebrandt. 1997. Determination of microbial diversity in environmental samples: pitfalls of PCR-based rRNA analysis. *FEMS Microbiology Reviews,* 21(3):213-229.

Wagner, M., M. Horn and H. Daims. 2003. Fluorescence in situ hybridisation for the identification and characterisation of prokaryotes. *Current Opinion in Microbiology,* 6(3):302-309.

Warnecke, F., R. Amann and J. Pernthaler. 2004. Actinobacterial 16S rRNA genes from freshwater habitats cluster in four distinct lineages. *Environmental Microbiology,* 6(3):242-253.

WHO. 2003. *Heterotrophic Plate Counts and Drinking-water Safety*. London, UK. IWA Publishing.

Wilczak, A., J. G. Jacangelo, J. P. Marcinko, L. H. Odell, G. J. Kirmeyer and R. L. Wolfe. 1996. Occurrence of nitrification in chloraminated distribution systems. *Journal American Water Works Association,* 88(7):74-85.

Williams, M. M., J. W. S. Domingo and M. C. Meckes. 2005. Population diversity in model potable water biofilms receiving chlorine or chloramine residual. *Biofouling,* 21(5-6):279-288.

Williams, M. M., J. W. S. Domingo, M. C. Meckes, C. A. Kelty and H. S. Rochon. 2004. Phylogenetic diversity of drinking water bacteria in a distribution system simulator. *Journal of Applied Microbiology,* 96(5):954-964.

Woese, C. R. 1987. Bacterial Evolution. *Microbiological Reviews,* 51(2):221-271.

Wolfe, R. L., N. I. Lieu, G. Izaguirre and E. G. Means. 1990. Ammonia-Oxidizing Bacteria in a Chloraminated Distribution-System - Seasonal Occurrence, Distribution, and Disinfection Resistance. *Applied and Environmental Microbiology,* 56(2):451-462.

Yang, J., G. W. Harrington, D. R. Noguera and K. K. Fleming. 2007. Risk analysis of nitrification occurrence in pilot-scale chloraminated distribution systems. *Journal of Water Supply Research and Technology-Aqua,* 56(5):293-311.

Yannarell, A. C. and E. W. Triplett. 2004. Within- and between-lake variability in the composition of bacterioplankton communities: investigations using multiple spatial scales. *Applied and Environmental Microbiology,* 70(1):214-223.

Zengler, K., G. Toledo, M. Rappe, J. Elkins, E. J. Mathur, J. M. Short and M. Keller. 2002. Cultivating the uncultured. *Proceedings of the National Academy of Sciences of the United States of America,* 99(24):15681-15686.

Zhen, D., H. Liu, S. J. Wang, J. J. Zhang, F. Zhao and N. Y. Zhou. 2006. Plasmid-mediated degradation of 4-chloronitrobenzene by newly isolated Pseudomonas putida strain ZWL73. *Applied Microbiology and Biotechnology,* 72(4):797-803.

ABBREVIATIONS

AOB	Ammonia-oxidizing bacteria
ARISA	Automated ribosomal intergenic spacer analysis
BDOC	Biodegradable dissolved organic carbon
bp	Base pairs
Chl-	Bench-scale chemostat not receiving chloramines
Chl+	Bench-scale chemostat receiving chloramines
DNA	Deoxyribonucleic acid
DO	Dissolved oxygen
DS	Distribution system
HI	High level of confidence matching an ARISA peak with microbial identity
HPC	Heterotrophic plate counts
HRT	Hydraulic residence time
ITS	Internal transcribed spacer
LO1	Pilot plant #1 operated at the Louisville Water Company
LO3	Pilot plant #3 operated at the Louisville Water Company
LOW	Low level of confidence matching an ARISA peak with microbial identity
MDN	Medium level of confidence matching an ARISA peak with microbial identity
MDS	Multidimensional scaling
MPN	Most probable number
N/A	Not available
NOB	Nitrite-oxidizing bacteria
OTU	Operational taxonomic unit
PCR	Polymerase chain reaction
R	Spearman's rank correlation
RDP	Ribosomal database project
RNA	Ribonucleic acid
rRNA	Ribosomal RNA
T	Type strain
USEPA	United States Environmental Protection Agency
UW	University of Wisconsin

UWH	Pilot-scale distribution system with high chloramine dose
UWL	Pilot-scale distribution system with low chloramine dose
2D	Two dimensional

IWA Publishing's authorised EU representative for General Product Safety Regulations is Diane D'Arras, 15 rue Duret, 75116 Paris, France, e-mail: safety@iwap.co.uk.

Printed and bound by CPI Group (UK) Ltd, Croydon, CR0 4YY
17/09/2026
02218813-0015